Arithmetic for Human Beings

ROBERT FROMAN

Simon and Schuster / New York

to Marilyn Marlow

Published by Simon and Schuster
Rockefeller Center, 630 Fifth Avenue
New York, New York 10020

SBN 671-21617-1
Library of Congress Catalog Card Number: 73-16476
Designed by Irving Perkins
Manufactured in the United States of America
1 2 3 4 5 6 7 8 9 10

Contents

1. The Crimes of the Textbooks

LABORING OVER ARITHMETIC

Millions of people loathe arithmetic. They have good reason to. The subject is taught in most elementary schools as if the whole idea is to make it loathsome. Only those born with great aptitude for it are likely to retain interest in it. This is just as true of the New Math kind of arithmetic as of the traditional kind.

But unlike the algebra and geometry inflicted on high-school students, arithmetic will not go away and leave you alone after you are paroled from the schoolroom. All life long it keeps popping up in supermarkets, kitchens, workshops, offices, almost everywhere—even in the far reaches of the backwoods. When it does, the schools have left most of us no alternative to going into a trance and trying to remember elementary textbook procedures. Some victims of New Math have more trouble at such times than those who endured only the traditional torments, but the difference is not fundamental.

Shocking though the news will be to all who have suffered from textbook arithmetic, there is a quite different approach much better suited to everyday use. It involves no tortuous shortcuts nor the memorizing of any formulas or systems. It is simple, easy, highly efficient, and not quite respectable. Most schools make sure that children learn nothing of it. This book seems to be the first attempt to reveal the naughty secrets in detail even to adults.

The revelation, it is to be hoped, will help bring to an end the reign of the sort of thing taught in the third and most revoltingly cute of the three Rs. But its end seems to be drawing near anyway. Although it has been treated as if it were eternal and immutable, it actually has held sway only for a few centuries. The adoption of at least a few splashes of New Math by more than two out of five American school systems demonstrates that dissatisfaction with traditional arithmetic runs deep, for New Math isn't a blissful solution. It makes as little sense to most school board members and teachers as to most parents and children. In theory an attempt to train children to do arithmetic in a rigorously deductive way, it is in fact a confusing change made mostly for the sake of change.

A vivid demonstration of how traditional arithmetic is inculcated was staged over a period of several years in the first decade of this century in the courtyard of an apartment house in Berlin. The chief actors were a retired schoolmaster, Herr Professor August von Osten, and his retired trotting horse, Hans. At about noon on pleasant days, any who cared to do so could drop in at the courtyard, where Hans was stabled, and watch him perform. His specialty was arithmetic. With taps of his right forefoot he provided answers to a variety of problems in addition, subtraction, multiplication, and division. He could perform these operations even with fractions.

Asked to multiply two-thirds by four-sevenths, for instance, Hans would first slowly tap eight times to indicate the numerator of the answer. Then he would tap twenty-one times to indicate the denominator.

There were plenty of skeptics. They tried again and again to catch von Osten at some kind of cheating, even though he never sought to make money from his horse's performances. Eventually, the horse—dubbed Clever Hans by newspaper feature writers—got to be such a sensation that a committee of psychologists, zoologists, and other experts investigated him. They enthusiastically authenticated his accomplishments. They felt that the final proof was their demonstration that Hans could solve arithmetic problems even in von Osten's absence, the problems being put to him and his answers observed by members of the committee.

Enter Oskar Pfungst, a young psychologist from the University of Berlin. He was not impressed by the verdict of this committee of his elders. In the fall of 1905 he launched his own investigation. It took him several months.

At first he produced only further corroboration of Hans's abilities. He found that even when a questioner whispered a problem in Hans's ear the horse could produce the correct answer. But finally Pfungst had the inspiration to have one person whisper a number into the horse's ear and another person, ignorant of the first number, whisper a command to add to it another number. This left the horse helpless. Even if the first whispered number was two and the command was "Add two," Hans could not perform the operation. He also failed when his eyes were covered and when it was too dark for him to see his questioner clearly.

Pfungst had the clues he needed, and he soon also had the explanation of Hans's prowess. The horse liked to please people and win their praise plus an occasional carrot or lump of sugar. He had discovered that he could do this in certain situations by tapping with his right forefoot and by stopping the tapping when the person he sought to please smiled, relaxed slightly, or made some other slight change in expression or posture. Indeed, Hans had become so expert at detecting the most minute of such signs that Pfungst found that he himself could not help giving a sign Hans could detect even when he knew the horse was looking for such a sign.

The point of this story is that horses are not alone in liking to please people and win praise. Most children share this taste. Many millions of them have been trained to add, subtract, multiply and divide in the same uncomprehending way as Hans. There also are available for use on children means of persuasion that were not used on Hans—threats, canings, labeling as a "failure," and for many the ultimate horror of "You'll never get into a college."

To be sure, few teachers of arithmetic consciously and deliberately exploit their victims. They are merely doing unto others as was done unto them in their own childhood. But it is time to break the vicious circle.

Consider an operation like multiplying 979 by 743. This would have been beyond Hans. The horse quite sensibly lost interest if he had to go on tapping that forefoot too long. But hundreds of millions of people have been led to believe that the operation can produce one—and only one—correct result and that there is a long and involved correct way of obtaining that result, although different school systems teach different correct ways.

A typical "correct" procedure is multiply the 979 by 3 and set down this result, then multiply the 979 by 40 and set down this result, then multiply the 979 by 700 and set down this result. Then you add these three results. The sum is supposed to be the one and only answer. The child who produces it is "right." The child whose answer is off by so much as a single unit is "wrong."

This is absurd, and within a generation or so probably will be recognized as such in most schools. All the various correct procedures are not only as mechanical as Hans's but also endlessly dull and tedious. Further, they provide many chances for mistakes. In the case cited there are more than forty chances for mistakes in calculating and in writing down the results of the calculations.

Children in many schools have it even worse than this. Teachers in such schools use computing problems as punishments, assigning them by the dozen to whisperers, fidgeters, gum chewers, and other such juvenile delinquents. Some teachers feel that the proper treatment for pupils who have trouble computing is to assign them great numbers of extra computing problems. A little worse is the insistence by some teachers that no shortcuts are permitted, even if they produce the "correct" answer. Some hapless teachers have been at this sort of thing so long that they have come to believe that computing drills are what mathematics is about.

These observations are not fresh off the top of anyone's head. They were among the findings of a study made by W. G. Quast, professor of elementary school mathematical education at Slippery Rock State College in Pennsylvania. He reported them in an article, "On

Computation and Drill," published in the December 1969 issue of *The Arithmetic Teacher,* a journal sponsored by the National Council of Teachers of Mathematics.

In other words, overuse and misuse of computing drill is an open scandal.

(Incidentally, the words "calculate" and "compute" will be used interchangeably in this book. A subtle distinction can be made between them, but it seldom is, and is not to the point here. The words "number" and "numeral" also will be used interchangeably, as is usual in everyday speech. To mathematicians, and to those who parrot certain New Math incantations, a number is "a property of a set" and a numeral is a symbol for a number, but the distinction is meaningful only in the higher reaches of mathematics and is of no concern to us here.)

A sizable minority of arithmetic teachers are battling for a change to more humane ways of introducing children to their subject, but the odds still are strongly against them. Many school administrators and school-board members who resist the New Math fad do so because they are outraged at any suggestion that there can have been something wrong with the way they, themselves, were taught arithmetic. Most arithmetic textbooks are written to please such administrators and boards, who are the buyers of such books. New Math textbooks also include a great deal of drill. It takes daring and inspired teachers to stand up against such forces.

Many of those who insist on computing drills for children are hypnotized by the indisputable fact that it is necessary to learn how to calculate in order to understand mathematics. They fail to distinguish between learning how and learning to be expert. It is necessary to perform only a few calculations in the process of learning how they are done. Lengthy drills are of use only to those who want to become expert.

The question is whether there is any reason why anyone today should want to become expert at calculating if it does not happen to appeal to him. The answer is a resounding *"No."*

Expertness at calculating is not necessary even to professional mathematicians. There are many stories about mathematicians who have been unable even to keep track of their own bank balances, and by no means all such stories are apocryphal. And although it is true that a few famous men of mathematics have been unusually adept at calculating, none has been able to perform feats comparable to those of the true calculating prodigies.

Many of these are called idiot savants, meaning that in matters other than the calculations in which they specialize they seem mentally retarded or at least quite dull. Others are of good general intelligence though unable to perform in other things at anywhere near the genius level of their calculations. Among the latter was a nineteenth-century Vermonter named Truman Safford. Although he grew up to be an astronomer and always was able to calculate rapidly, it was only in his childhood that he performed prodigies.

At the age of ten Safford was asked to multiply in his head the number 365,365,365,365,365,365 by itself. An observer reported the boy's response to this request:

"He flew around the room like a top, pulled his pantaloons over the tops of his boots, bit his hands, rolled his eyes in their sockets, sometimes smiling and talking, and then seeming to be in an agony, until, in not more than one minute, said he, 133,491,850,208,566,925,-016,658,299,941,583,255."

As remarked by James R. Newman, who repeated this story in his anthology *The World of Mathematics:*

"An electronic computer might do the job a little faster, but it wouldn't be as much fun to watch."

The only case in which it makes sense to ask a human being to perform long and involved calculations is when he is one of those rare prodigies who can put on such a show. Asking an ordinary human being to do such work is equivalent to asking him to prepare a wheat field for planting by turning over the earth with a spading fork. Tractors and gang plows do not only a far faster but also a far better job. Calculating machines compute more quickly and more accurately than any human being can hope to.

Note the term "calculating machine." So much has been said and written recently about the great electronic computers developed since World War II that the simpler devices tend to be forgotten. Some arithmetic teachers, for instance, concede that the development of computers makes it possible that eventually there will be no need to drill pupils to make them proficient, but most such teachers tend to place that eventuality in a distant future.

In fact, various simpler devices have been doing most of our calculating for several generations. There was a time when store clerks, for instance, had to be quick and accurate, at least at addition. That time is long gone. The cash register began taking over the work more than a century ago, and today almost all the arithmetic a clerk absolutely has to have is the ability to read numerals.

Another, simpler type of calculating device far antedates the cash register and still competes with it in many parts of the world. This is the abacus. Mesopotamian, Egyptian, Greek, Roman and other ancient ways of writing numerals made arithmetic operations extremely difficult. Almost all calculations had to be done with abacus-like devices. In fact, the word "calculate" comes from the Latin word for the pebbles used in one kind of abacus.

It was some unknown Hindu who, a century or two before or after 500 A.D., invented a symbol for zero and made possible the positional system of notation. This is the system we use today in which the position of a symbol in a numeral determines the symbol's value. In 11, for instance, the 1 on the left stands for one ten and the 1 on the right stands for one unit. In 10 the 1 on the left also stands for one ten and the 0 on the right stands for no units. Until the invention of a symbol to stand for no units (or no tens as in 103, or no hundreds as in 1037, and so on) this kind of notation was not possible.

Ironically, the inventor of the symbol for zero apparently meant it to stand simply for an empty column in an abacus (abacuses being arranged in columns standing for units, tens, hundreds, etc.). What makes it ironical is that it was this invention that eventually made possible the abandonment of the abacus in favor of what has become our traditional style of arithmetic. Only with the positional system of notation is it possible for children to perform such arithmetic operations without the help of an abacus.

But the positional system of notation merely made it *possible* for children to perform such operations. That system did not make it necessary or desirable for either children or adults to become so proficient at computation that they could take over all such work previously done on the abacus. It is true that expert human computers can perform some traditional arithmetic operations faster than these can be done with an abacus, but other such operations can be done much faster and more accurately with that device.

Snobbery seems to have been the most important of the various reasons for western Europe's abandonment of the abacus beginning about four hundred years ago. That device was worked with the hands, and a true gentleman did not work with his hands. A true gentleman sent his children to schools where they could be caned repeatedly until they memorized the multiplication table and other mechanical ways of producing "correct" answers.

But this book is not a plea for a return to the abacus. Nor is it concerned with anything so dull as the simple-minded procedures used in the operation of more modern calculating machines. Abacus arithmetic, calculating-machine arithmetic, and third-R arithmetic are all basically the same. What concerns us here is a kind of arithmetic beyond the powers of any machine, and suitable only for human use.

In fact, the kind of arithmetic offered here can help adults free themselves from powerful unconscious pressures to behave like machines. It is possible for a human being to choose between or make a blend of two different ways of coping with any situation that arises. One is the logical, analytical way, the cause-and-effect way, the "if this, then that" way. It involves being as precise as possible about causes in order to try to be reasonably clear about effects.

In some situations this way is highly successful all by itself. But some devotees are so devoted to it that they are incessantly anxious about "getting things straight," "lining up the facts," "straight thinking," and so on. Such straightness can become a mental straitjacket. It perverts logical thinking by attempting to make it do everything when it can properly and successfully do only some things.

The other way of coping is by trying to see the situation whole or, at least, to encompass large segments of it in gulps, patterns, *gestalts.* This is what leads to hunches and inspired guesses. It usually is called intuition.

Those who distrust or even suppress this kind of thinking and rely wholly on the analytical approach may be literally short-circuiting their minds, even their physical brains. Many different lines of research have established that the two halves of the human brain specialize in different functions. One specialization, for instance, is in the control of the right side of the body by the left half of the brain in right-handers. The left half of the brain also seems to dominate in the analytical thinking of right-handers, while the right half dominates in intuitive thinking. (For left-handers the functions of the two halves of the brain seem to be reversed.)

It also is possible, of course, to overvalue intuition to the detriment of analytical thinking. Some people operate mostly by guesses and, though their guesses prove illogically correct on occasion, they usually need insulation from reality in the form of money or neurosis to survive their many mistakes. There is even one kind of arithmetic, to be discussed in a later chapter, "The Percentages," in which intuition leads many people astray in some situations. But the essence of the harm done by textbook arithmetic is its unremitting insistence that attention be focused exclusively on the seemingly neat, logical steps of the arithmetic operations.

Some interesting discoveries of how this short-circuits

the mind were made by the late Max Wertheimer, a leading psychologist and student of the learning process. In his pioneering book, *Productive Thinking,* he reported on the results of a variety of studies indicating the debilitating effects of textbook arithmetic. He summed up the essence of his findings about this in a quotation from a fellow psychologist to whom he had just outlined some of his results:

"Of course, I see. It reminds me of an experience I had some months ago which may be typical. My son, who is a bright boy, came to me and said, 'You see, Daddy, I am very good in arithmetic in school. I can do addition, subtraction, multiplication, division, anything you like, very quickly and without mistakes. The trouble is, often I don't know *which* of them to use.'"

What makes this especially horrifying is that in everyday life it is almost impossible to short-circuit the mind to this extent. Such a thorough job of confusion can be done only on children who can be made to feel that what the textbooks serve up is unquestionable truth. When you encounter an actual problem in the solution of which arithmetic can be helpful, it almost always is obvious which of the arithmetic operations is to be performed.

That schoolroom experience of arithmetic is much like what continues to short-circuit many adult minds in later years. For such adults that experience has made unquestionable the belief that in any arithmetic question the numbers involved are all-important and the proper way to perform each of the four basic operations invariable.

In real life the numbers are *not* the most important parts of problems. What the numbers stand for is what matters most. How the necessary operations on those numbers should be performed depends on what you want to know. To decide this, you have to see the problem whole, not in neatly numbered pieces.

There are several different ways to go about each of the arithmetic operations, and it is impossible to lay down a law about which is the best way in all situations. Textbooks pretend to do just that, even those whose compilers have the best and most honest intentions. The intentions of the compilers of textbooks are not always the best, or even honest.

Morris Kline, professor of mathematics at New York University's Courant Institute of Mathematical Sciences, went into detail on the subject of textbooks in his recently published *Why Johnny Can't Add,* a devastating analysis of some of the effects of the New Math. Kline described how textbook compilers sometimes plagiarize each other quite openly, paraphrasing page after page and whole chapters. He asked one fellow professor of mathematics why he had put together a certain typical textbook.

The reply: "Oh, I can write the stuff between classes without even having to think about it. Why shouldn't I?"

Arithmetic for Human Beings is intended as the opposite of a textbook, a kind of anti-textbook. It is primarily a book to be read, not labored over. The suggestions it offers are not going to be repeated endlessly, summarized at the ends of chapters, and reinforced with large numbers of space-filling examples and exercises. Such paraphernalia would be useless for the purpose here. The only way you can learn to make use of these suggestions is by trying them in situations that matter to you, not on artificial problems dreamed up by someone else. The suggestions are meaningful only when you really want to know something that arithmetic can help you find out.

So let us begin. And to emphasize that the going is to be far different than in most books dealing with arithmetic, we shall begin not with addition, as is usual, but with multiplication. This is the operation most dramatically different here from its equivalent in the textbook style of arithmetic.

2. Making "Times" Make Sense

MULTIPLICATION

There are several ways to respond to an apparent need, request, or challenge to perform such an operation as multiplying 979 by 743. To get an idea of some of the possibilities, consider how a person might respond to the question "How old are you?"

If the one questioned happens to be in his or her forties or older, he or she might well respond with only a withering look or with a rasping "None of your business" or with a bland "Twenty-nine." All these are not only legitimate responses but ones a questioner ought to have the sense to anticipate. They or others like them also are legitimate responses to a challenge to multiply any pair of numbers if you are given no idea of what the numbers stand for. To be sure, there are many circumstances in which it probably would be best to try to find out what they are supposed to stand for. But the important thing is to break the habit of responding in the old, drilled way.

There also are a variety of ways in which young people might respond to the "How old are you?" question. Suppose that the questionee is a boy whose last birthday was several months back, and suppose that that birthday was his tenth.

He could truthfully answer, "Ten."

Or he could truthfully answer, "Going on eleven."

He could also take a little time and work out the answer, "Ten years, seven months, and seventeen days." He might even add something like ". . . and six hours." And if he did this, he probably would evoke in his questioner the pained look of one who is forced to realize that he is dealing with a youthful smart aleck.

Anyone seeking such precise information concerning the age of another person would not ask, "How old are you?" He would ask something like "Do you know how old you are to the day or hour?"

The question "How old are you?" really means "About how old are you?" And the answer "Ten" really means "About ten."

That word *about* is an important tool in the kind of arithmetic this book is concerned with. Some other important tools are the words *almost* and *nearly* and the phrases *a little more than* and *a little less than.*

It almost always is a waste of time to seek a precise answer to any arithmetic problem that comes up in everyday life. Such an answer can even be misleading because it distracts attention from what really matters in the situation at hand. Those who always insist on such answers are suffering an affliction similar to that of a boy who insists on stating his age to the day or hour.

What all these considerations have to do with multiplying 979 by 743 is this:

You can see at a glance that 979 is about one thousand.

And 743 thousands are 743,000. Therefore, 743 times 979 is about 743,000.

It's that simple. And that easy. And often, that's all there is to it.

That's even more than there need be to it if, for instance, you are trying to decide whether a certain pilot of supersonic jet planes is eligible for the "Million Mile Club." Suppose you have the information that the average speed of the planes he has flown has been 743 miles per hour and that he has flown them for 979 hours. In that case you might scarcely notice even that 743,000 figure. You would just say to yourself something like:

"Hmmm. A little less than three-quarters of a million miles. Still quite a way to go to make the club."

This is the correct answer to your question. A more precise product of the two factors 979 and 743 (numbers to be multiplied together are called factors and the result of the operation is called the product) would not be correct because it would be a waste of time.

One reason why this approach to arithmetic is considered disreputable and unfit for children is that schools treat arithmetic as the first step on the ladder to higher mathematics. In those lofty realms it is necessary to be precise about arithmetic operations because the numbers and other concepts dealt with are abstractions. The connections with reality often are tenuous or obscure. Consequently, you cannot be sure that every last digit will not turn out to be of great importance in whatever you are so deviously getting at.

This helps explain the textbook approach but does not justify it. Higher mathematics is indispensable in the functioning of the modern world. Some people find it a fascinating subject, even esthetically stirring and satisfying. But these are not excuses for burdening everyone with the concept of arithmetic essential in such pursuits while hiding away the techniques most useful in the arithmetic of everyday life.

Do not get the impression, however, that what's being offered here is so-called mental arithmetic. Whether or not you write down any numbers you are operating with depends entirely on you. If you are more comfortable with written numbers, by all means write them. If you like to work with numbers in your head, work with them that way. The important thing is to relax and take it easy—so try any approach that helps you do this.

To get back to 743 times 979—there are, of course, many cases in which you would have good reason for wanting a more precise result than "a little less than three-quarters of a million." It's up to you to decide this, however. How precise you want to be depends on what the numbers stand for and your interest in what they stand for.

Suppose you are a retailer and have orders for 979 gadgets on which you can make a profit of $7.43 each. Unfortunately, you cannot order exactly 979 of them from your supplier. In fact, he will ship the gadgets to you, at the price enabling you to make that $7.43 profit on each of them, only in lots of one hundred dozen.

You must buy, that is, 1200 of the gadgets, or 200 more than a thousand. Your orders for 979 come to about 20 less than a thousand. So you will have about 220 of them left after filling the order.

In your area these gadgets are hard to sell. Your orders for 979 of them are the results of some extraordinary circumstances. It is almost certain that you will have to sell some of the extra two hundred-odd at or below cost, and you cannot ignore the possibility, though it is a rather remote one, that you may be unable to sell any of them at any price.

Each gadget, including handling charges and a properly budgeted share of overhead, will cost you about $30. So your possible greatest loss is 220 times $30. This should be an easy one to work out, but if it is not easy for you, think of the $30 as three $10s. 220 times $10 is $2200. Three times $2200 is $6600. (If this still gives you any trouble, take the result on faith here. Ways of handling those zeroes will be discussed a little later.)

The question is—are you justified in risking this loss of $6600 in order to make a profit of 979 times $7.43?

You can think of $7.43 as 743 cents. As we already have seen, 743 times 979 is about 743,000. To change 743,000 cents into dollars, all you have to do is drop those last two zeroes. So the profit to be made comes to about $7430.

It can be seen at a glance that $7430 is several hundred more than $6600, but the amounts are not terribly far apart. This is the reason for wanting in this case to be a little more precise concerning the results of multiplying 979 by 743. For since 979 is less than a thousand, 979 times $7.43 is less than $7430.

But it is easy to be as much more precise as there is any reason to be. As we have seen, 979 is about 20 less than a thousand. All that's needed is to deduct 20 times $7.43 from $7430. Indeed, you can forget the 43 cents and concentrate on the $7 which, when multiplied by 20, yields an easy $140. Deducting this from $7430 leaves nearly $7300—still obviously several hundred dollars more than that maximum loss risk of $6600. (Actually, it may not be necessary even to bother deducting the $140 from the $7430. You know that the difference between $6600 and $7430 is a great deal more than $140, and this may be all that matters.)

It has taken several paragraphs to go through this process of working out and comparing these figures. Once you get the hang of handling figures this way you can do such things in just a few moments. In this particular case you would be able to tell at a glance that the assured profit is a few hundred over $7000 and the maximum possible loss a few hundred less than that figure.

But the gain in speed is by no means the only advantage to this kind of arithmetic. In many cases it is a minor consideration. These are the cases in which the fact that the figures produced by this approach are imprecise helps to clarify your thinking about the matter in hand. The case we have been considering is of that sort.

When you calculate 979 times $7.43 in the automatic, textbook way, either with pencil and paper or with the help of a machine, and calculate all the other figures involved here in the same way, the investment of time and effort is sizable. Such an investment creates value. It makes you feel that these figures are valuable and meaningful in themselves and distracts your attention from the extreme tentativeness of the figure for the maximum loss risked. If for some reason the assured profit of a few hundred dollars seems less than enough to make the deal definitely worthwhile, the tentativeness of the maximum-loss figure is the heart of the matter. All attention should go to that figure and the possibility of reducing it. Directing attention that way is distinctly easier and more natural when the arithmetic that produces the figure is, itself, tentative.

To put it another way, the textbook approach to such an arithmetic question short-circuits your mind. It makes you concentrate on the analytical thinking involved in the multiplication and prevents you from using your intuition. What you need most in this case is an inspiration about how to dispose of the extra gadgets at a good price.

At this point you may be beginning to feel a bit restless. Multiplying 979 by 743, you may reasonably think, has been pretty well taken care of. How about other cases—especially cases in which neither of the figures to be multiplied is close to 1000 or some other nice, round number?

The answer is that any pair of numbers to be multiplied can be fiddled with until both are nice, round numbers.

Suppose you want to multiply 63 by 44. Fiddle with the 63 and make it 60, the nearest round number. Treat the 44 similarly, making it 40.

Now, 60 is six tens, and 40 is four tens. Ten times ten is one hundred. Four times six is twenty-four. So 40 times 60 is twenty-four hundred or 2400.

This is a good way off the mark, though there often is no reason to come any closer. But in the event that you do want to get closer, all it takes to get there is a little more fiddling. Before going into how this second fiddling is done, however, it is necessary to deal with a few matters that have been slid over a bit sneakily.

One such slide is the assertion that multiplying two round numbers, such as 40 times 60, is a simple operation. For many people it is simple. But for others more deeply scarred by the traditional way of teaching arithmetic, there are stumbling blocks.

Almost everyone can readily see that ten times ten is one hundred. But having got this figure, some have trouble combining it with the twenty-four produced by multiplying six by four. They have trouble, that is, deciding whether the result of multiplying 60 by 40 is 2400 or 240.

Here are a couple of tips on getting around this difficulty—one for the eye-minded and one for the ear-minded. If you like to visualize the numbers with which you are working, it will help to concentrate on the zeroes at the end of the number. Remember that whenever you multiply round numbers you wind up with at least as many zeroes at the end of the product as you start with at the ends of the two numbers you are multiplying.

In multiplying 60 by 40, for instance, you wind up with two zeroes at the end of the product—2400. If you were multiplying 600 by 40, you would wind up with three zeroes at the end of the product—24,000. And 400 times 600 results in a product ending in four zeroes—240,000. You can wind up with more zeroes than you start with (40 times 50 gives a product of 2000, for instance, since the product of 4 times 5 also ends in a zero), but never with fewer.

For the ear-minded the suggestion is to concentrate on the names of the round numbers. Forty times sixty means, as we have seen, four tens times six tens, and ten times ten is one hundred. Once you know this you know that your final product is going to be in hundreds, and the product of four times six tells you just how many hundreds.

Another stumbling block is that old nemesis of generations of bored elementary school pupils—the multiplication table. Some of us reacted to the endless efforts to make us memorize the table by getting it stuck in our minds so firmly that we never can encounter a challenge to multiply, for instance, six by four, without instantly producing the answer twenty-four. But others, our betters in resistance to tyranny, either valiantly refused to submit from the very beginning or else ejected the whole thing from their minds as soon as they escaped the schoolroom.

Such resisters not only fought the good fight. They also missed nothing of great importance. Even those among them who find in later life that they have occasion to do a good many multiplication chores without the help of machines can get along without memorizing the table. All they need do is make a chart like this:

1	2							
2	4	3						
3	6	9	4					
4	8	12	16	5				
5	10	15	20	25	6			
6	12	18	24	30	36	7		
7	14	21	28	35	42	49	8	
8	16	24	32	40	48	56	64	9
9	18	27	36	45	54	63	72	81

With the help of a couple of pencils or strips of cardboard this makes the table instantly available. Just put one pencil under the line begun on the left by the larger of the numbers to be multiplied and the other pencil on the right side of the column headed by the other number. The product is in the corner where the two pencils meet.

This, for instance, is the way it works with six times four:

1	2							
2	4	3						
3	6	9	4					
4	8	12	16	5				
5	10	15	20	25	6			
6	12	18	(24)	30	36	7		
7	14	21	28	35	42	49	8	
8	16	24	32	40	48	56	64	9
9	18	27	36	45	54	63	72	81

Any who have occasion to use such a chart much will soon find that they have memorized the table painlessly, which is the sensible way to do it at any time of life. You cannot learn anything well unless you want to, though some of us can be coerced into parroting certain answers to certain questions. But once a person wants to learn something, it comes easy.

Notice that the table goes only to nine times nine. In some schools it has been the custom to try to force children to memorize the multiplication table up to twelve times twelve or even to twenty times twenty. There is no harm in such memorizing for those who enjoy it. There is no harm in memorizing a table up to one hundred times one hundred if it pleases you to do so. But it is cruel to impose demands for such memorization on the unwilling, because there is little use for a table beyond the nine-times-nine level.

To multiply by ten simply means to add a zero to the end of the number multiplied. And to multiply by a number a little higher than ten, all that's necessary is to put a zero at the end of the number being multiplied, then add to this product the result of multiplying by the amount by which the multiplier exceeds ten. For example, to multiply 12 by 11 you can simply put a zero at the end of 12, thus producing 120, then add to 120 the result of multiplying 12 by 1.

In some New Math classrooms a great deal is made of what is called the distributive law that multiplying by a number taken apart produces the same result as multiplying with it whole. That 10 times 12 plus 1 times 12 produces the same result as 11 times 12 is a case of this. The supposed justification for the fuss is that there are branches of higher mathematics in which this does not hold true. Such exceptions are far from the fields of everyday arithmetic.

If the New Math chatter or anything else makes you uneasy about whether this "law" holds, try both ways with a few examples. Try whether 2 times 2 produces the same result as 1 times 2 plus 1 times 2. Or whether 10 times 10 produces the same result as 5 times 10 plus 5 times 10, or 3 times 10 plus 7 times 10.

Convince yourself. Taking another person's word for it is almost sure to lead to confusion eventually.

Another New Math fuss is about what is called the commutative law—that it does not matter in what order you multiply numbers. Again, this usually is obvious in everyday arithmetic. It means that 2 times 3 is the same as 3 times 2. Or that 11 times 12 is the same as 12 times

11. Or that 44 times 63 is the same as 63 times 44. If you ever have any doubts, try both ways.

A third such consideration is the associative law—that when you want to multiply together three or more numbers you can do it in any order you find convenient. If you want to multiply 2 times 3 times 4, you can first multiply 2 times 3 to get 6, then 6 times 4. Or you can multiply 2 times 4 to get 8, then 8 times 3. Or you can multiply 3 times 4 to get 12, then 12 times 2.

Like the other two laws, this should be obvious, but it is easily testable if doubts ever assail you.

Now we can get back to fiddling with numbers in order to make them easy to deal with in multiplication. The instance being considered a few pages back was the multiplication of 63 by 44. It was suggested that each of these be changed to the nearest round number, thus making the task the simpler one of multiplying 60 by 40, which easily yields 2400.

You know that this is somewhat off the mark since 60 is less than 63 and 40 is less than 44. The question is whether it is so far off as to matter. Whether or not it is depends on what the 63 and 44 stand for.

Suppose that what's involved is the question whether you have the means to transport a group of people on a trip. There are available 44 buses with capacity for 63 passengers each. The number of people to be transported is about 2000.

In this case that figure of 2400 is all you need. It tells you that there is more than enough room for all who may want to go on the trip. Since 60 is less than 63 and 40 less than 44, the product of the round numbers obviously is less than the product of the numbers they are substituting for.

But if the number of people to be transported is about 2800, matters are quite different. In this case it is essential to know about how far off the mark that 2400 figure is. This is where the second fiddling comes in. And now what gets fiddled with are the round numbers.

Instead of making the 63 a plain 60, you can think of it as 60 plus 3.

You can similarly think of the 44 as 40 plus 4.

And you can think of 44 times 63 as 40 plus 4 times 60 plus 3.

This means:

40 times 60
plus 4 times 60
plus 3 times 40
plus 3 times 4

In the act of fiddling this way with any pair of numbers to be multiplied you arrange them so that they, themselves, emphasize what matters to you.

The arrangement shows that the first pair of numbers to be multiplied are going to yield by far the most important part of the product. Indeed, since 40 is ten times 4, it is obvious that 40 times 60 yields a product ten times the size of 4 times 60. The other two products are smaller still.

If you visualize the numbers clearly, you proceed with the next steps in your head. If not, you jot them down. In either case what you see is:

40 times 60 yields 2400
4 times 60 yields 200 plus 40
3 times 40 yields 100 plus 20
3 times 4 is not worth bothering about.

Then you see that 2400 plus 200 plus 100 comes to 2700. And adding 40 plus 20 leaves you a bit short of 2800.

This tells you what you want to know about whether 44 buses with a capacity of 63 passengers each are enough to transport about 2800 people. It tells you that the most important part of the figure about 2800 is the about. If you can be fairly sure that somewhat fewer than 2800 will show up, you know that you probably can get by. If not, you know that you may have either to squeeze more passengers into each bus or try to arrange for at least one more bus or leave a few late-comers behind.

This is the imprecise way things almost always work out in actual situations. An operation like 44 times 63 looks like it is going to produce a definite, clear-cut result. But when you examine what the figures stand for, you find that one or all of them are approximations.

Sometimes the nearest round number is a little higher rather than lower as in the cases of the two numbers we have been considering. You could round off 67, for instance, to 70. And if you needed more detail about the result of multiplying by 67, you would treat it as $70 - 3$. (If multiplying by a negative number gives you any trouble, a way to approach the operation will be considered in Chapter 7.)

People who are compulsively neat and tidy may have trouble accepting this way of handling numbers. For many such persons arithmetic is a refuge where neatness and tidiness are inexorable. It is part of a sort of dream world of exact sciences and clean-cut and immutable facts.

In reality, an important part of the exactness of the exact sciences is clear recognition that there are strict limits to the precision of the facts being dealt with. One of the forms that recognition takes is an approach to arithmetic something like the approach being set forth here. Given two figures to be multiplied one by the other, for instance, research workers do not do anything, no matter how swift a computer they have at hand, before determining which is the least precise of the two. The degree of precision of the least precise of two factors determines the degree of precision of their product.

Suppose you want to know the area of one surface of a rectangular strip of metal. And suppose you are informed that it is 2.0 inches wide and 3.22 inches long. It may seem obvious that all you have to do is multiply 3.22 by 2 and feel sure that the area is exactly 6.44 square inches. But this is not the case. All you can be sure of is that the area is about 6.4 square inches.

This is because the figure for the width is 2.0 inches—not 2.00 inches. That is, the number 2.0 does not provide any information about what comes after the 0. Consequently, you have to take the figure 2.0 inches as meaning somewhere between 1.96 inches and 2.04 inches, because both 1.96 and 2.04 round off to 2.0 as do all other figures between them (such as 1.96444, 1.999, 2.000001, and 2.04444). If you want a more precise figure for the width, you can try to measure more precisely, but there are no "absolutely precise" or "perfectly exact" measurements.

The rounding-off process is so important in the way scientists deal with figures that they usually adopt a strict rule about what to do when the figure to be dropped is exactly 5—that is, when they want to round off a number like 2.5 or 3.5 or 9375 or 9365 by dropping the last digit. The rule is to add 1 to the preceding digit half the time; the other half the 5 is simply dropped and the preceding digit is left unchanged. One way to accomplish this is by adding 1 when the preceding digit is even and dropping the 5 without changing the preceding digit when the latter is odd. In the cases cited above, the 2.5 would become 3, the 3.5 also would become 3, the 9375 would become 9370 and the 9365 also would become 9370.

Whenever you want to multiply one number by another, then, the first thing to do is relax. Or as athletes put it, stay loose. Shake yourself free of all those school years of terror about whether your answer would be "correct." Remember that the product you want is the one that will best suit your purposes.

If one of the factors is a single digit and the other only two digits long, the product you want probably is staring you in the eye. With practice you will be able to recognize it with little more effort than it takes to recognize an old friend.

7 times 83, for instance, is about 7 times 8 times 10.

As you can either remember automatically or quickly find on your chart, 7 times 8 is 56. And 10 times 56 is 560.

It is likely that is all you need to know. But if there is reason to be more precise, all you have to do is add to 560 the product of 7 times 3.

Indeed, when you gain experience in this approach, you will find it almost as easy as this to do any multiplication in which one of the factors is only a single digit, even if the other factor runs to four or five or more digits.

Suppose you want to transport six packing cases full of something or other, and you know that each of them weighs 3419 pounds. You have available for the job a truck with a stated capacity of ten tons. Your question is whether the truck is capable of carrying all six cases in one load.

A ton is 2000 pounds, so ten tons come to 20,000 pounds. What you want to know is how close 6 times 3419 comes to 20,000.

You can easily see that 6 times 3000 is 18,000. Just as easily you can see that 6 times 400 is 2400. The figure 6 times 19 does not matter much, but you can think of it as about 6 times 20 or a little over 100.

18,000 plus 2000 comes to 20,000, so the six packing cases will exceed 20,000 pounds by 400 plus about 100 pounds.

This focuses your attention where it should go—on the question whether the figure given as the truck's carrying capacity is elastic enough to permit adding a little more than 500 pounds.

To recapitulate—when you want to multiply two numbers, take them apart and focus first on the product of the biggest parts. If that does not tell you all you need to know, add the products of other parts beginning with the next biggest and proceeding on down the size line until you can see that the remaining products will make no difference that matters to you.

Above all—remember that you are not seeking the "correct" answer. You are trying to find out something you want to know, and you alone are the judge of whether or not you have found it.

3. The Easy Summer

ADDITION

Those interested in the history of mathematics have thought long and hard about how the human habit of adding things up may have gotten started. Some scholars have felt sure that the likeliest place to look for clues is in the languages of primitive tribes. Among the simplest approaches to adding that have yet been found was that of the now extinct aboriginal inhabitants of Tasmania, the island off the southeastern tip of Australia. The Tasmanians were a shy, gentle, not very numerous people, and the nineteenth-century European invaders of their island found it convenient to kill all of them. But before the last Tasmanians died, anthropologists managed to find out a little about their ways of living and thinking, one such finding being that the Tasmanian system of adding things up consisted in its entirety of "one, two, plenty." That is to say, their only number higher than two was plenty. Many of the products of Western civilization considered this laughable, proof that the Tasmanians were scarcely human and therefore worthy of the oblivion they were helped to.

It is true, of course, that having available only the label *plenty* for any collection of more than two things would be inconvenient in many situations. But the Tasmanian way of adding also has its advantages, especially over the procedure of addition inflicted on children in the textbook approach. For if the Tasmanian way gets to plenty in something of a hurry for many reasonable human purposes, what of a way of handling numbers that postpones endlessly and needlessly the arrival at plenty?

The Third R approach to addition, which insists that one keep track of every last penny, millimeter, ounce, second, or whatever, is a powerful reinforcement of the neurosis of affluence. It takes for granted—and thereby helps make respectable—the most abysmal greed. It is a truly Satanic device for turning what should be the simplest and easiest of arithmetic operations into an anxious, miserly ordeal.

Near the heart of the matter is the story everyone has heard about how a local bank once kept its employees working deep into the night in order to account for a single missing penny. Bankers being at least as gullible as non-bankers when it comes to myths about money, it is more than likely that this sort of thing has happened in some banks. More rational bankers no doubt make excuses for such waste of time and money by imagining it a kind of advertising for their "integrity" and a way of making people think that banks do not tolerate mistakes. The chief actual effect seems to have been to reinforce in most bankers the belief that they must never admit making arithmetic mistakes of so much as a single penny.

Presumably the inflexible and simple-minded logic of the big electronic computers eventually will make bankers aware that occasional minor mistakes must be, and easily can be, taken in stride, and that it is insulting to their customers to insist on pretending otherwise. Even the computers make mistakes. What the rest of us have to learn in order to approach addition sensibly is more difficult. We have to learn to break a habit not only taught us by arithmetic teachers but reinforced by countless repetitions of such incantations as "A penny saved is a penny earned" and "Take care of the pennies and the dollars will take care of themselves." We have to learn to ignore the pennies. Even more difficult, not to say heretical—we often need to ignore the dimes and dollars, too.

A typical everyday question involving addition goes something like this: Your records show that on the first of the month you had in your checking account $304.48. They also show that since the first you have written checks for $102.17, $40.61, $32.56, $18.00, $6.15. You receive a bill for $50.00. Your question is whether you can pay this bill without danger of overdrawing your account.

The textbook way of tackling this problem is to write down one under another the amounts of the checks already drawn:

```
$102.17
  40.61
  32.56
  18.00
   6.15
-------
```

You start with the column on the right, adding 5 and 6 and 1 and 7, write 9 under the line at the bottom of the column, carry 1, add 1 and 1 and 5 and 6 and 1—and so on.

Like the textbook approach to multiplication, this is not only a compulsive waste of time but also an endless teetering on the brink of mistakes that can falsify the result. All you really need do is note, first, that $102.17 is about one hundred dollars. Then you note that $40.61 is about forty dollars and $32.56 is about thirty dollars and $18.00 plus $6.15 is about twenty dollars, and that forty plus thirty plus twenty comes to about another hundred dollars. You thus see that you have written checks for about one hundred and about another hundred. One, two, leaves plenty out of the three available hundreds to cover a check for another $50. And you are right in there with the primitive Tasmanians—which is to say way ahead of the overcivilized unfortunates who have to do it the textbook way or not at all.

Just as in multiplication, the first step is to overcome the effect of having been led during childhood to take it for granted—to "know" without ever thinking about the matter—that any question involving addition is a problem like the problems in textbooks and that such a problem has one and only one correct answer. In most situations it is utterly irrelevant to turn a question involving addition into a textbook style of problem. In the case under consideration doing so would produce an answer in dollars and cents. What's wanted is a simple yes or no.

Unfortunately, there is a catch for those most grievously hurt by the way arithmetic has been taught in schools. The catch is in seeing that the five checks written so far come to about one hundred and about another hundred. That $102.17 is about one hundred should be easy enough, but a considerable number of victims of the textbooks are bound to have trouble with the reasoning that "$40.61 is about forty dollars and $32.56 is about thirty dollars and $18.00 plus $6.15 is about twenty dollars and that forty plus thirty plus twenty comes to about another hundred dollars."

The first step in getting past this trouble is to be quite clear about the effects of ignoring the pennies, dimes, and dollars in these figures. Consider the pennies and dimes first. There are none of either in the $18.00 figure, so you can skip it. But suppose that there were the maximum number of both in the other three figures—suppose, that is, that these figures were $40.99, $32.99, and $6.99. Three times 99 cents is about three dollars. But even three dollars would make no difference in the answer to the question under consideration, because if the sum of the checks you have written were two hundred and three dollars, there still would be plenty left out of the three hundred to cover another fifty-dollar check. So you can certainly ignore the three actual figures of 61 cents, 56 cents, and 5 cents, which obviously total much less than three dollars.

Next, consider the effect of ignoring the dollars. Suppose that the $32.56 figure were $39 and that the sum of $18.00 and $6.15 were $29. Two times $9 can be thought of as about $20. But even an extra twenty dollars would make no difference in the answer to the question under consideration, because if the sum of the checks you have written were two hundred and twenty dollars, there still would be plenty left out of three hundred to cover another fifty-dollar check. So you can certainly ignore the actual $2 in the $32.56 figure and the figure for the number of dollars by which $18.00 plus $6.15 exceeds twenty dollars. Indeed, there is no reason even to bother working out the figure for the number of dollars by which $18.00 plus $6.15 exceeds twenty dollars, since it obviously is less than nine dollars.

Finally, there is the reasoning that "forty plus thirty

plus twenty comes to about another hundred." Many people find it easy to work this out either mentally or by jotting the figures down. But those most valiant in defiance of the hickory stick may have trouble with it, especially if the defiance or their natural inclinations have led them to adopt the habit of doing their adding on their fingers.

Even the defiant ones tend to be a little shamefaced about this practice, because teachers sometimes go to great lengths in efforts to make it seem shameful. This is another of the absurdities of the standard approach to arithmetic. Some people seem to have an inborn need for objects they can feel or manipulate in the process of counting or adding if they are to perform these operations with confidence. But whether the habit develops out of innate tendencies or as an expression of defiance, there is no need to try to break it. One of the advantages of the approach to addition being suggested here is that it makes expedient the use of fingers even when the numbers being added run to tens, hundreds, thousands, or greater sizes.

Adding forty dollars plus thirty dollars plus twenty dollars is easy if you like to do your adding that way. Just think of these amounts as three piles of ten-dollar bills. There are four of the bills in one pile, three in another, and two in the last. So all you really have to do is count off four plus three plus two fingers. It would be the same if you were adding hundreds, thousands, tens of thousands, or whatever.

This is not to suggest, however, that you *should* add the four plus three plus two on your fingers. Add them that way only if it seems natural and easy to you. Otherwise, add them in your head or with pencil and paper or in any other fashion you like.

In whatever way you do it, what you are doing is performing a feat of memory. You know that one and one makes two because you long ago memorized this "fact." It is more meaningful to call it a convention; one and one make two only because everyone agrees to call one and one not "one and one" but "two." The convention probably got started because one and one takes three words and two only one word. One and one and one takes five words, two and one takes three words, three takes only one word, and as the size of the collections of ones grows, the advantage of a new name for each collection also grows.

There is a great difference between the way most of us become able to perform the feats of memory involved in counting, or adding, and the way we learn, or refuse to learn, the multiplication table. The latter usually is shoved at us in school. Addition is dangled irresistibly before us by our peers or slightly older friends long before we are sentenced to the schoolroom. We learn how to do it not because we are told to but because we want to.

Unfortunately, teachers often try to force children to forget what they have spontaneously learned about adding and to go about it some other way. This is true not only in those cases in which the spontaneously learned way involves use of the fingers or other such aids to memory. It also is likely to happen when children develop for themselves any aids to doing addition their own way.

Probably the best aid, and the best way to break loose from the textbook approach to addition, is to think of it as simply a way of rearranging numbers for greater convenience. Just as it usually is inconvenient to think "one and one and one" rather than "three," so is it usually inconvenient to think of four and three and two as "four and three and two" rather than as "nine."

The steps by which you go about rearranging four and three and two into nine also are a matter of suiting your convenience. Different people find different ways more convenient. Most would do it in two stages, though one might get through the first stage so rapidly as to be only half-conscious of having done so. One such first stage is combining the four and three to produce seven, the second stage being the adding of two to seven. Some would add the three and two first, then add the four to the resultant five. A few might prefer to add the four and two, then the three. Another few might take one from the four, reducing it to three, add this one to the two, raising it to three, then combine the three threes. The procedure that suits you best is the correct one for you.

It becomes more difficult to rearrange numbers conveniently as the numbers get bigger, and a sharp change in the rearranging procedure occurs after nine. Just as you know—because you long ago memorized the convention—that eight plus one is nine, so you know that nine plus one is ten. But there is a great difference between nine and ten, much greater than the difference between nine and eight. The difference is made visible in the way we write numerals.

9 is an arbitrary symbol purposely made quite distinct from its predecessors 0, 1, 2, 3, 4, 5, 6, 7, and 8. On the other hand, 10 simply combines the first two symbols. It is agreed that the 1 in 10 stands for one ten, which is a number consisting of ten ones, or units. The 0 in ten means that there are no extra units in addition to that set of ten. The first 1 in 11 means the same as the 1

in 10, and the second 1 in 11 means that there is one extra unit. The 2 in 12 stands for two extra units in addition to the set of ten. And so on.

Why do it this way? It obviously would be possible to invent a different name and symbol for each number. Nine and one could be called *donk,* and the symbol ȼ could stand for donk. Donk and one could be called *mimp* and written ᚷ. Mimp and one could be called *leent* and written ᴗ. And so on.

These names and signs are every bit as natural and reasonable—which is to say as artificial and arbitrary—as the names and signs for the numbers zero to nine. The trouble with them is that it would take a great deal of time and effort to memorize a different name and sign for each number up to even such a comparatively low one as what we now call one hundred. Hardly anyone would know them all up to what we call one thousand. And any kind of operation with such numbers—adding mimp and leent, for instance—would involve another feat of memory. It is far more convenient to start over again at one every time you reach ten.

That this is what we do is slightly obscured by the terms eleven and twelve. Because of the way the English language has changed over the centuries, these names seem almost as arbitrary as mimp and leent. The immediate ancestor of the modern word eleven was the Middle English *enleven,* which was a contraction of a phrase meaning "one left over." Twelve is a descendant of a phrase meaning "two left over." Both names leave obscure that they imply "*ten and* one (or two) left over."

Thirteen is a little clearer in this respect. It is easy to see ten in -teen, and if you change thir- back to thri-, it comes recognizably close to three. That it means "ten-and-three-left-over" while thirty means "three-tens" is a little arbitrary, but easy to go along with once you see that this is the intention. It also is easy enough to see or hear two-tens in twenty, four-tens in forty, and so on. That one hundred means ten tens is arbitrary, but the symbol 100 is clearly derived from 10. The same goes for 1000, 10,000 and so on.

But the most important result of accepting convenience as a proper goal of arithmetic operations is that it can be a big help in shaking off the textbook habits. It is not absolutely obligatory to rearrange nine-and-two into ten-and-one; no one is going to whack you with a hickory stick if you don't do it that way. The sole reason for the ten-and-one arrangement is that it is convenient *in some cases.* In other cases it may be more convenient either to leave the nine-and-two alone or to rearrange it some other way. In fact, there are many cases in which the arrangement "about ten" is more convenient than ten-and-one.

This is so not in spite of, but because of, the fact that "about ten" is in many situations a convenient way to think of many other numbers, such as eight, seven-and-two, eight-and-four, nine, eleven, and so on. If what you are trying to decide is whether there is plenty of such and such on hand, and if about ten is plenty, then it is beside the point and distracting to think of any of these numbers in any way other than as about ten.

The same, of course, applies to rearranging forty plus thirty plus twenty as about one hundred, and this brings us back to the business of deciding whether there is plenty for certain specific purposes in your bank account. To recapitulate—you had in your account on the first of the month $304.48. You already had written checks for $102.17, $40.61, $32.56, $18.00, and $6.15. You know that this means you have written checks for about one hundred and about another hundred, which leaves plenty to make it possible for you to pay a bill for $50. So you pay it.

Suppose you then receive a bill for $63.74. Do you have enough left in your account to pay it?

This is still a yes-or-no question, not one requiring an answer in dollars and cents right down to and including the last hard-squeezed penny. But it takes a little more time than the first question. You know that you have written checks for about one hundred and about another hundred plus fifty. This makes it clear that an additional check for $63.74 will bring you quite close to that $304.48 mark, so you need to know more concerning those "abouts."

The answer to the first question came so easily that there was no need to inspect closely the figures involved—but now look at them more closely. Do you notice something about the $32.56 and the $18.00? What's to be noticed is that 32 and 18 can be neatly rearranged as 50. (Some will see this at a glance; others arrive at it in stages, noting first perhaps that the 8 and 2 can be thought of as 10, that 30 and 10 can be thought of as 40, and that adding the other 10 left over from the 18 raises the 40 to 50.) This 50 and the 50 of the new $50.00 check just written easily rearrange as 100, with only the 56¢ in the $32.56 being ignored.

Something similar is to be noticed about the $40.61 and the new bill for $63.74. The 40 and the 60 also easily rearrange as 100. So now there are three 100s—those two we have just created and the one in the $102.17 check. They clearly amount to a little more than $300. And there is still that check for $6.15.

So you have your answer. No, you cannot write a check for $63.74 unless you make a new deposit or can depend on your bank to hold still for a small overdraft until you get around to making a deposit.

You can make this approach to bankbook balancing still easier by ignoring the pennies and dimes in your deposits and withdrawals. In keeping track of them just round off each one to the nearest dollar. This means, of course, that you never will know to the penny how much you have in your account. But is it worth knowing?

It does not often happen in everyday life that you want to know anything about the sum of a large number of numbers, but such occasions do arise. For instance, you may be zipping up and down the aisles of a busy supermarket, happily piling goodies in your shopping cart, then suddenly realize that you have only a twenty-dollar bill in your pocket. You don't have a checkbook with you, and there is no hope of getting the management to let you use a blank check. The only thing to do is keep your purchases under the $20 mark.

There's the basket already piled with seventeen items. The prices read .19, .59, .43, 1.27, .93, .65, 2.42, .67, .10, .38, .89, 2.25, 3.85, .13, .75, 1.10, .26. You do not even have pencil and paper. Where to begin?

This is the sort of situation that beautifully demonstrates what's wrong with the textbook approach. If it were the only approach possible, most of us would be helpless. Probably not one person in a hundred can mentally add so many numbers the textbook way. But that is far from the best way to get the answer sought.

First, pick out and put by themselves all the items costing a dollar or more. There are five of them, their prices being $3.85, $2.42, $2.25, $1.27, and $1.10. $3.85 is close to $4, so call it $4. $4 and $2 is $6. And $2 is $8. And $1 is $9. And $1 is $10. The 42 cents in the $2.42 is about 50 cents, and 25 cents and 27 cents make about another 50 cents, and the two 50 cents make $1. Add this to the $10, and your total so far is $11.

There are twelve other items. Just by looking them over you can see that they range from 10 cents to 93 cents and that most of the 10-cent intervals in between these two are represented (that is, there is one item in the teens, one in the twenties, one in the thirties, and so on). If you take their average value as 50 cents, you will not be far off. Twelve 50-cent items would come to $6. $6 and $11 come to $17. You can feel confident that you are a good $2 or more below your $20 limit.

It will not always happen in such situations, of course, that the large number of small figures will be scattered over the entire range from near zero to near 100 (or from 10 cents to near $1, as in this case). The twelve lower-priced items in your basket might have prices like these: .19, .59, .63, .93, .65, .67, .10, .78, .89, .23, .75, and .26. But here again you can see just by examining them that four items fall in the range from 10 cents to 26 cents and eight fall in the 59 cents to 93 cents range. You also can see just by looking that of the eight in the upper range five are at 75 cents or below. So it will be quite safe to treat the eight items as averaging 75 cents apiece. Again, there are many ways to work out that eight 75-cent items total $6. The other four items average less than 20 cents apiece and so total less than $1. Since $11 and $6 and $1 come to only $18, you still are well under your $20 limit.

It also might happen that all twelve items priced less than $1 would be in the upper range and average about 75¢ apiece. This would work out to a total of $9 for these twelve items, and adding that to the $11 total of the over $1 items comes to $20 for the whole contents of your basket. In that case you know that it would behoove you to hold back one or two of the items of least importance to you until you see how the cash register works out the total of the more important ones.

This approach to the addition of a large number of numbers often makes it unnecessary to bother at all with the smaller items. Suppose you are ferrying building materials from a little fishing port to an offshore island. You know that your boat can carry two and one-half tons, or 5000 pounds, in addition to you and your helper and still ride safely high above the water in any except the very roughest weather. Another five hundred pounds will be safe in very calm weather, but this is your absolute limit. You must order for delivery at the dock and ready for loading onto your boat only what you can carry because of the danger of theft or vandalization of anything left unattended on the dock while you are gone. You have a list of the weights in pounds of the items to be carried. This list reads: 811, 13, 10.2, 24, 290, 17.6, 1476, 73, 21.7, 64.2, 179, 8.3, 12.9, 517, 320, 84, 14.3, 9.2, 4.1, 66, 479, 5.5, 14.6, 782, 3.7, 19, 72.5, 185, 62.9, 6.3, 18, 79, 214.

Glancing over this you can pick out three different kinds of figures. One kind is in the hundreds of pounds. A second kind is in the 60s-to-80s range. The third kind is in the range from less than 10 pounds to a little over 20 pounds.

Obviously, the first kind is the one to consider first. You can write them down separately. Or you can go down the list and encircle each number over 100. While you are at it, you might as well underline the numbers in the 60 to 80 range for later checking:

(811), 13, 10.2, 24, (290), 17.6, (1476), 73, 21.7, 64.2, (179), 8.3, 12.9, (517), (320), 84, 14.3, 9.2, 4.1, 66, (479), 5.5, 14.6, (782), 3.7, 19, 72.5, (185), 62.9, 6.3, 18, 79, (214).

Start with the first of the over 100s—namely 811. Call it 800. The next one is 290. Call it 300. 800 and 300 rearrange easily as 1100. Plus 1500 (i.e., 1476) makes 2600. Plus 200 (179) makes 2800. Plus 500 (517) makes 3300. Plus 300 (320) makes 3600. Plus 500 (479) makes 4100. Plus 800 (782) makes 4900. Plus 200 (185) makes 5100. Plus 200 (214) makes 5300.

This is close to your absolute limit. You can see that there are seven items in the 60 to 80 group and that they probably average more than 70 pounds each, and that seven times 70 is 490 or about 500 pounds. This makes it obvious that you cannot take everything in one trip. There is no need to bother about the other figures.

All of which may look a bit long-winded the way it has been spread over these pages. The appearance is deceptive; only the explanations take time and space. The actual operations are far quicker and more efficient even than using an adding machine. Here you focus your attention on what matters most instead of trying to do a merely mechanical job of adding numbers to whose meaning you pay little attention. This leaves you free to concentrate on the human specialty of thinking and making a decision.

It also leaves you free to consider new possibilities. When you quickly discover that the load is too big for one trip, you have more time to consider such a possibility as taking most of it in the first trip and trying to find other cargo for inclusion in the second load.

Another advantage to the suggested approach is that it cuts down the chances that you will make a big and fateful mistake. If you were to write all the figures down one under another the way you were taught to do in elementary school—or if you were to punch them all on the keys of an adding machine—you probably would give exactly the same amount of attention to the smallest item—the 4.1—as to the biggest—the 1476. What's more, in starting with the right-hand column you find a lot of figures to add, and they seem big and important. That single 1 in the left hand column—the 1 in 1476—would be all too easy to overlook. Such a mistake would mean either a dangerously overloaded boat, a lot of time wasted in carting materials back to the lumberyard, or a number of items left on the dock at the mercy of any thieves or vandals who might chance by during your absence. By concentrating on the most important items first you minimize the possibility of overlooking any of them.

4. "Neither a Borrower Nor a Lender Be"

SUBTRACTION

In a rigorously informal survey of attitudes toward the subtracting operation seven persons were asked whether any one aspect of the operation stood out as especially bothersome for them. Six of the seven supplied the same answer. They did so almost instantaneously and with vehemence:

"Borrowing!"

The seventh questionee took more time and words:

"It's taking one from the next number so that you can raise the one you are subtracting from high enough to make it bigger than the one you are subtracting from it—that's what fouls me up every time."

This is borrowing by another and decidedly long-winded name. It has many names. Sometimes it is called "the take-away-and-carry-method." Sometimes "the borrowing-and-repaying plan." Sometimes "decompositions." Sometimes "the additive method of subtraction."

These and other terms, some of which involve slight

differences in procedure, seem to have been invented in efforts to ease the difficulties so many people have with borrowing. Most of the terms and procedural variations are in use today, for the method for subtraction never has been quite standardized. Even the compilers of textbooks have been forced to recognize that their approaches to the operation have drawbacks. Yet they seem never to have been able to bring themselves to consider the sensible way out of their difficulty.

By whatever name borrowing be known, it makes no sense if it is not helpful. It is a device intended to ease the operation of subtraction, not an immutable law of nature, man, or God. If it bothers you, make crude waxen images of the teachers who inflicted it on you, stick pins in the images, and throw them in a fire. Or do whatever else is necessary to rid yourself of the handicap of believing that in arithmetic it is necessary to flout Polonius' advice to Laertes:

> Neither a borrower nor a lender be;
> For loan oft loses both itself and friend,
> And borrowing dulls the edge of husbandry.

An example of the kind of operation in which borrowing is usually supposed to be essential is subtracting nine from thirty-three. In the most common textbook way of approaching this undertaking you write the numbers like this:

$$\begin{array}{r} 33 \\ \underline{9} \end{array}$$

This is supposed to make clear that the thirty-three consists of three tens and three ones and the nine of nine ones; and that since nine is greater than three, you must "borrow" one set of ten ones from the three tens, reducing the three to two, add the ten borrowed ones to the three ones to produce thirteen ones, then subtract the nine ones from the thirteen ones.

This is as verbose as poor old Polonius and lacks any hint of the cracks of wisdom that lard his doddering foolishness. The sensible way to subtract nine from thirty-three is by thinking of nine as about ten. One ten from three tens leaves two tens, so your answer is about twenty-three.

If you have reason to be more precise, you know that nine actually is one less than ten, so you have subtracted one too many. Add that one to your preliminary result, and you have your precise answer—twenty-four.

This, obviously, is a very simple example. Subtracting, say, fourteen from thirty-three may present more difficulties for some. And subtracting 837 from 2614 or $46.79 from $531.18 may seem to be in another "higher" category of operation. But all these subtractions are better done without recourse to borrowing.

Before getting down to specifics, it will be helpful and interesting to consider the nature of the subtraction operation. Its very name and the old-fashioned names for the various parts of the operation are suggestive of a worried and reluctant approach to a disagreeable undertaking. To add numbers can be pleasant and profitable. But to subtract is to cut, to take away, to whittle down, to learn the worst. To be sure, there are occasions when what you whittle down is a debt you owe and are making payments on, or a debt owed you on which you are receiving payments. But the atmosphere of loss predominates, so it is no wonder that the operation causes uneasiness.

Textbooks used to thicken the gloom by inflicting on generation after generation of sad-faced children the necessity of memorizing and keeping straight the terms *minuend, subtrahend,* and *remainder* for the numbers party to the operation. These sound like ingredients of a spell to be recited in the more garishly lugubrious style of funeral parlor. If you ever stumble on any of them lurking in dark corners of your mind, run them out. Even the compilers of textbooks seem to have shaken loose from them in recent years.

What some New Math texts use in place of these terms does not improve the situation. If condemned to their approach, you might find yourself yearning for the dear old funeral-parlor stuff. One New Math definition of "the operation called subtraction" takes several pages studded with gems like: "The family belonging to a difference $a - b$ consists of all those differences $x - y$ for which $a + y = x + b$."

Although it may be hard to believe, this sort of thing actually makes sense in a kind of defining necessary to some branches of mathematics. It does not make sense in the kind of subtracting being considered here. All that concerns us in this chapter is taking the Little One from the Big One to find What's Left. (Cases in which you take the Big One from the Little One and wind up with less than nothing will be considered in a later chapter on negative numbers.)

Another unfortunate aspect of the textbook approach to subtraction is the assertion in many such works that the operation is just the reverse or inverse of addition. The implications of this superficially true statement are extremely misleading. The two operations differ in fundamental ways.

For one thing, two plus three and three plus two both

produce the same sum. But you cannot switch around like that in subtraction. Three minus two has a result quite different from two minus three.

Also, you can add three pencils to two pens and wind up with five items, but you cannot subtract two pens from three pencils. If all you have are pencils and you want to give someone pens, you need the connivance of a third party.

Another important difference between addition and subtraction is that you can add a long list of numbers in one operation, but each subtraction is necessarily a separate operation. If you want to subtract two and three and four from ten, you must do it serially instead of all at once. That is, you take two from ten, then three from what's left, then four from what's left after the second operation. Or you must first add the two plus three plus four, then subtract this from ten.

All of which is obvious to most people most of the time. It is worth putting into words here only because the obvious has a way of getting slippery in your grasp once in a while if you take it too much for granted. A more commonly helpful way of comparing in your own mind and in daily life the operations of adding and subtracting is to think of subtracting as a more sharply focused kind of rearranging of numbers. In adding five and three, for instance, you can think of five plus three as just that and keep the five and the three separate indefinitely while you consider how to rearrange them or while you do something else. In other words, you are interested in all the numbers equally and can postpone as long as you wish the process of rearranging the five and three as eight or as about ten or as six-and-two or as four-and-four or in whatever other way suits your need or whim.

In subtracting three from five you can, of course, postpone the operation as long as you wish. But once you perform it, your interest, like that of a surgeon, shifts from the part taken away to what's left. Your concern is for the patient, not the tissue removed.

The secret of how to avoid the stumbling block called borrowing is to keep firmly in mind this preeminent concern for the patient—for what's left after the operation. You can do almost anything you wish to the tissue to be removed—certainly anything that will make it easier for you and the patient. And in some cases you may want to prepare the patient for the operation by shaving off any superficial growth that might get in your way.

Spelling the numbers out or thinking of them in words rather than symbols—that is, thinking of one rather than 1, two rather than 2, and so on—is a good way to prepare both patient and tissue to be removed for this new way of operating. It also can help you, the surgeon, to break operating habits inculcated in the schoolroom. So, to get back to the matter of subtracting a number of one figure from a number of two figures, consider taking one, two, three, four, five, six, seven, or eight from thirty-three. This is, of course, a very minor operation, and for just that reason excellent for practice purposes.

Subtracting one, two, or three from thirty-three ought to seem the same as subtracting one, two, or three from three. If it seems somehow different, remember that the term thirty-three is simply a short way of writing or thinking "thirty plus three." If all that you want to take away is one, two, or three, all you operate on is the three.

One way of preparing such tissue as four or five or six for subtracting from thirty-three is by thinking of the four or five or six as about three. Taking three from thirty-three leaves thirty. Taking about three from thirty-three leaves about thirty.

One way to prepare seven or eight for subtracting from thirty-three is the same as the way used earlier in subtracting nine from thirty-three. Just think of seven or eight as about ten. About ten from thirty-three leaves about twenty-three.

If the numbers stand for certain kinds of things, this sort of information is all you can hope for. Suppose you are driving from Centerville to Outertown and the road map has indicated a distance of thirty-three miles between them. If you pass a sign pointing back the way you have come and reading "One (or two, three, four, five, or six) miles to Centerville" and if what you want to know is how far you still have to go to Outertown, all you can be sure of is that you still have about thirty miles to go. And if the sign indicates that you are seven, eight, or nine miles from Centerville, all you can be sure of is that you have about twenty-three miles to go. Highway maps and road signs are notoriously approximate in their distance figures.

But on other occasions you have more exact information and need a more precise result. Suppose you are in charge of a group of thirty-three children and have sent some of them on errands. You want to take all those left in a group from wherever you are to some other place, so you must make sure that all of them are with you.

If those on errands number one, two, or three, no borrowing is involved even in the textbook approach to the problem. If you have sent off four or more of the children, you can avoid borrowing by preparing the tissue to be removed a little differently than in the highway miles case. Just snip a neat, precise three off both the tissue and the patient, reducing the patient to a neat precise thirty and reducing the tissue to be removed by a neat precise three. Usually, it is easy to subtract what is left—that is, to see at a glance if the number of children on errands is five, that three from five leaves two more to be subtracted from thirty and that thirty minus two leaves twenty-eight. But if taking two or more from thirty gives you any trouble, count down on your fingers. One down is twenty-nine, two down is twenty-eight, three down is twenty-seven, and so on.

If the number of children is seven, eight, or nine, you may still want to proceed in this same way, counting down four, five, or six from thirty. Or you may prefer to lop ten off thirty-three, thus reducing it to twenty-three and ruining the comparison with careful surgical procedure (which is a good idea because that metaphor has done all it can for us and is ready for retirement). Now you have subtracted too much, but it is easy to see exactly how much too much by subtracting the seven, eight, or nine from ten. Then you add to twenty-three the result of this subtraction—that is, you add three to twenty-three if you are subtracting seven from thirty-three, two if you are subtracting eight, or one if you are subtracting nine.

All of which has taken a lot of words—which makes it seem complicated. It actually is quite simple and can be stated succinctly. To subtract a number of one figure from a number of two (or more) figures, first use all you need of the Little One to reduce the Big One to a round number if the Little One is big enough to do this. Then take away from the round number whatever is left of the Little One. Or if you prefer, lop ten off the Big One and add back the result of subtracting the Little One from ten.

Subtracting a Little One of two figures from a Big One

of two or more figures is more complicated but not a great deal so. At its simplest, such as in subtracting eleven from thirty-three, it involves no borrowing even in the textbook method. But in that method you first take one from the three, reducing it to two, then take one ten from the three tens, reducing them to two tens and leaving you with two tens plus two, or twenty-two. This is the mechanical way.

In the human way you start at the other end—the important end, the "business end." You know that eleven is about ten and that about ten from thirty-three leaves about twenty-three, which may be all you need to know. Only if you decide you need to know more do you bother noticing that eleven is one more than ten and that you must take one more from twenty-three to get an exact answer.

It is very easy, of course, to subtract eleven from thirty-three, and when the operation is written out in the textbook way as 33 − 11 or as

$$\begin{array}{r} 33 \\ \underline{11}, \end{array}$$

many people can see at a glance that what's to be done is to take 1 from each 3. But those who have great trouble with borrowing sometimes find difficult such a closely similar operation as taking fourteen from thirty-three. There are two different flexibly human ways to approach an operation like this.

If—and only if—you happen to notice that taking thirteen from thirty-three would reduce it to a nice round twenty, you have the beginning of the easiest way. That's because if you notice this you are almost certain to notice simultaneously that thirteen and fourteen are about the same, and this means that you know that taking fourteen from thirty-three leaves about twenty. In cases when you want more precise information, you know that fourteen is one more than thirteen and that this means you must take one more from twenty.

There are many subtraction operations in which this approach will work quickly and easily for most people once they get the hang of it. If you want to take twenty-seven from ninety-five, you may notice that twenty-five from ninety-five leaves a round seventy. If you want to take fifty-three from two hundred and sixty-one, you may notice that fifty from two hundred and sixty leaves a round two hundred and ten. If you want to take eighty-seven from four hundred and twenty-one, you may notice that one hundred from four hundred and twenty leaves a round three hundred and twenty.

The other way to approach operations like these—the way to use when you cannot see an obvious possibility like those suggested—is a little more formal, though by no means mechanical. To subtract fourteen from thirty-three you first take one ten from the three tens and then notice that taking away four more wipes out the three. To subtract twenty-seven from ninety-five you first take two tens from the nine tens and then notice that taking away seven more wipes out the five. To subtract fifty-three from two hundred and sixty-one you take five tens from six tens and notice that taking away three more wipes out the one. To subtract eighty-seven from four hundred and twenty-one you note that eighty-seven is about ninety which is nine tens which is about ten tens which is one hundred, then you take the one hundred from four hundred and twenty and are left with about three hundred and twenty.

In all of these operations you decide in advance how

precise you want your answer to be. If it must be exact, you keep track at each step—either in your head or with pencil and paper—of how much too much or how much too little you are subtracting. Then you either take away what remains to be subtracted or add back what you have oversubtracted.

In subtracting eighty-seven from four hundred and twenty-one, for instance, you note that eighty-seven is three less than ninety and nine tens are one ten fewer than ten tens. You have subtracted thirteen too much. Also, for convenience sake you dropped the one from four hundred and twenty-one. Add the thirteen and the one to the tentative answer of three hundred and twenty, and you have your precise answer of three hundred and thirty-four.

Again, this has taken many words to explain and sounds much more complicated than it is. The basic idea can be put in five words: go for the round numbers. Sometimes it is only the Little One that needs to be rounded off; sometimes the Big One too. And if you need a precise figure for What's Left, get it by making up for whatever you have to do in the rounding off process.

This brings us to operations involving more figures, such as subtracting 837 from 2614 or $46.79 from $531.18. As suggested earlier, these may seem to belong to another and "higher" category than we have been considering, but they are easily brought down to a comfortably human level. The way to do this works equally well on those rare occasions in daily life when you have to perform a subtraction operation on numbers even longer than these.

Suppose that in the case of the first pair of figures you are on a committee trying to make sure of the availability of hotel accommodations for a potential 2614 persons at a convention. You know that one large hotel has signed up to provide accommodations for 837. The question is—how many more remain to be found?

It certainly would be pointless—and possibly worse than pointless—to go about this operation the textbook way. If you even write down the numbers with the Little One under the Big One

$$\begin{array}{r} 2614 \\ \underline{837} \end{array},$$

you may trap yourself into the old classroom imitation of a brainless machine. The only numbers that make any sense at all here are the round ones. There probably is less than one chance in a thousand that exactly the potential 2614 will show up, and considerable chance that the hotel with room for 837 can accommodate at least a few more. To treat the two numbers as if they are meaningful down to their last digits (which is the way one treats all numbers in the textbook approach to arithmetic) may be worse than pointless in this case because it may make anyone else present who has a more human approach to numbers think you are foolishly fussy and should be kept as far as possible away from any matter of real importance.

This is another of the many kinds of occasions when the convenience of the numerals can hinder clear thinking. It is easier to write 2614 and 837 or to think "two-six-one-four" and "eight-three-seven" than to write or think "twenty-six hundred and fourteen" and "eight hundred and thirty-seven." But the numerals can hypnotize you and make each figure seem important, while the full name of a number calls attention to the more important part.

All you actually know here is that about twenty-six hundred persons may show up and that accommodations have been arranged for about eight hundred of them. The only meaningful operation is taking eight from twenty-six. When you gain a little experience with this approach, you will see at a glance that you still need to find space for about eighteen hundred and that this is the only useful information available from this particular pair of numbers.

If the twenty-six hundred and fourteen stands for the number of dollars of income tax you owe for the year and the eight hundred and thirty-seven for the number of dollars you have already paid, matters are quite different. Unless you are feeling rather remarkably, not to say astoundingly, generous toward the Internal Revenue Service, you will not be satisfied with a figure like "about eighteen hundred" as what you still owe. Neither, for that matter, will the I.R.S. Its computers cannot digest an unmechanized term like "about."

The reason for your own dissatisfaction should be that you can see:

First, that eight hundred from twenty-six hundred leaves eighteen hundred;

Second, that thirty-seven is bigger than fourteen;

Third, that this means that the precise figure is less than eighteen hundred.

The next step toward the precise figure you want is to note that you cannot take thirty-seven from fourteen (not, that is, without getting into negative numbers, which are not yet on the schedule). So you take the Little One from the Big One—i.e., fourteen from thirty-seven. What's Left is the amount by which the precise figure is less than eighteen hundred.

Taking fourteen from thirty-seven should be easy by now—one ten from three tens leaving two tens, or twenty, and four from seven leaving three. So the precise figure is twenty-three less than eighteen hundred. There are, as we have seen, two different flexibly human ways of going about an operation like subtracting twenty-three from eighteen hundred:

First, if you notice that twenty-three is close to twenty-five and realize that there are four twenty-fives in one hundred and that three of them make seventy-five, you will know at once that eighteen hundred minus twenty-three is about seventeen hundred and seventy-five. Since twenty-three actually is two less than twenty-five, subtracting twenty-five means subtracting two too much, so you add back two and get the precise answer of seventeen hundred and seventy-seven.

Second, if you do not notice an easy shortcut like this, think of eighteen hundred minus twenty-three as seventeen hundred plus What's Left when you take twenty-three from one hundred. Twenty from one hundred is two tens from ten tens, and What's Left is eight tens or eighty. That leaves three to subtract from eighty. If the result of that operation is not obvious to you, you can count down on your fingers—seventy-nine, seventy-eight, seventy-seven. Or you can think of eighty minus three as seventy plus What's Left when you take three from ten.

And as usual, you can work with any of these methods in your head or you can jot down whatever numbers you want to jot down. If you jot, you can either spell out the names of the numbers, or make up abbreviations that please you, or use the standard numerals. The idea is to please yourself and yourself alone. The thing to beware of is being tricked into following mechanical textbook procedures.

The other complicated-looking subtraction operation mentioned earlier was $531.18 – $46.79. These are the figures from an actual case, a sad case of textbook poisoning. A college student buying his first car saw an advertisement for a used one that sounded good to him. The advertised price was $449, and when he telephoned the dealer to confirm the price, he was assured that was it. The boy went to the lot, examined the car, drove it a few miles, and decided that he wanted to buy. He then learned that the dealer wanted $531.18 for it.

When the boy protested, the dealer blandly explained that this included the 5% sales tax and a few other small charges plus $46.79 for a new radio a mechanic had "mistakenly" installed between the time the boy had called about the advertisement and the time of his arrival to inspect the car. The boy said he did not want the radio. The dealer told him to try the car for a week or two with the radio. If at the end of that time he still did not want the radio, he could bring the car back, the radio would be removed, and his $46.79 cheerfully refunded.

"That," went on the dealer, handing the boy pencil and paper, "will bring the price down to what you seem to think is all you can afford. Figure it out for yourself."

Like millions of others, the boy knew that this kind of arithmetic was one of his weak points. But he was hypnotized, partly by the dealer's blandness and partly by the habit instilled by endless schoolroom drilling of automatic mechanical response to a question involving arithmetic. He wrote

$$\begin{array}{r} 531.18 \\ \underline{46.79} \end{array}$$

and started plugging away at the textbook subtraction operation. Before he could stumble all the way through it, the dealer was spieling again. The boy wound up driving out of the lot with the radio still in the car and with himself signed up to pay $531.18.

In a situation like this, a far better way for the boy to approach the operation $531.18 – $46.79 would have been by stepping back and considering the whole situation. A swindler's mode of operation is to keep his victim's attention away from what matters. Nothing is so distracting as doing arithmetic the textbook way. Although the boy returned to the lot again and again to have the radio removed and get back his $46.79, he was never able to find the dealer in.

But this was a minor part of the swindle. A worse result of being distracted by the absurd arithmetic problem was that the distraction prevented the boy from noticing that the dealer had picked him for a sucker and from wondering why. What could have made the dealer think him gullible enough to swallow the story about the "mistaken" installation of the radio? What, indeed, but the fact that the boy was willing to pay $449 for this particular car. It cost more than $300 in repairs to keep it running for three months, at the end of which he was able to sell it for only $75 and probably was lucky to get out from under even at that substantial loss.

In other situations in which you do need some idea of What's Left when $46.79 is subtracted from $531.18, the first step is to notice that $46.79 is about $50 and $531.18 is about $530. This means that the operation involves taking five ten-dollar bills from a pile of fifty-three ten-dollar bills. Taking away three of the five leaves fifty, and you can easily count down the two others from there—forty-nine, forty-eight. So one good result is forty-eight ten-dollar bills, or about $480.

If you need a more precise figure, notice that in this first approximation you have subtracted a little too much. $46.79 is about forty-seven, which is three less than fifty. In subtracting fifty you thus subtracted three too much. Add it to the $480 and get the closer approximation, $483. Then notice that $531 is one more than $530. Add this to $483 and get the still closer approximation, $484.

On those few occasions when there is reason to care about dimes and pennies, you notice that $46.79 amounts to two dimes and one penny less than the $47 you have subtracted. Add this 21¢ to $484 and you have $484.21. Now add to this the 18¢ of $531.18, and the rarely necessary precise answer is $484.39.

There is an easy way to check your result at any stage of approximation. What's Left plus the Little One should be the Big One. In the case we have been considering, the first approximation to What's Left was $480, and the Little One was $46.79. When you have a feel for this approach to arithmetic, you can quickly see that this sum is about $527. You also can see that $527 is about $4 less than the Big One—$531.18. This means that adding $4 to $480 will give you a close approximation.

As in the case of the other arithmetic operations, then, the first step in a properly human kind of subtraction operation is to consider what the numbers stand for and what you hope to find out by performing the operation. Then start with the important parts of the numbers.

Don't bother with the lesser parts unless you have reason to care about them. Filling out your income tax form might be one such reason.

There is, however, one frequently encountered special case in subtraction. When the Big One is very large and the Little One quite small, What's Left probably does not differ significantly from the Big One. In such cases, that is, the operation does not change the Big One in a way worth bothering about.

This is the secret of success of much petty thievery. Subtracting a few nails from a hundred-pound keg of nails or a couple of shots from a quart of whiskey two-thirds full does not make a change likely to be noticed. This often is the undoing of petty thieves. Success goes to their heads, they try to graduate to the big time, and wind up in the Big House.

Somewhat similar is the practice, which few merchants dare abandon, of pricing merchandise not at $1 but at 99 cents, not at $2000 but at $1995, and so on. This is almost universal even though sales taxes have rendered it meaningless in many cases, since, if a sales tax is involved, a 99-cent price tag turns out to mean an actual payment of $1.02 or thereabouts. Yet again and again buyers have stayed away in droves from items priced in round numbers, then started buying with abandon when the prices were lowered a penny or so.

This seems to be another of the results of the mechanical schoolroom approach to numbers. Mechanically there is a distinct and impressive difference between 99 cents and $1. It takes only a little thought to enable a person to realize that the difference is unimportant, but those who know only the mechanical approach often are blocked from doing any thinking at all about numbers.

5. The Guessing Game

DIVISION

In the year 944 A.D. a monk in an English monastery wanted to divide six thousand one hundred and fifty-two by fifteen. He went about it by writing down in Roman numerals all of the multiples of fifteen up to six thousand—that is, XV, XXX, XLV, LX, and so on. Then he counted these multiples. He found that there were four hundred of them and concluded that there are four hundred fifteens in six thousand. To take care of the remaining hundred and fifty-two, he went back over his list of multiples and found that it took ten of them to reach one hundred and fifty. He concluded that in six thousand one hundred and fifty-two there are four hundred and ten fifteens, with two ones left over.

This tale usually is told as evidence of how cumbersome the Roman numerals make the division operation. The idea is that the Hindu-Arabic numerals we use now make it much faster and easier. And indeed they can help make it very easy. But they seldom do so in textbooks or schoolrooms.

In the fifteenth century, when Italians had been using

the Hindu-Arabic numerals for a couple of centuries, a friar named Luca Pacioli, a writer on arithmetic, expressed the general opinion of the division operation in a line with a built-in sigh: "If a man can divide well, everything else is easy."

Things have changed little in this respect since Pacioli's time. In the October 1971 issue of *The Arithmetic Teacher,* an elementary school principal complained:

"Learning to work division problems is probably the most difficult arithmetic task for elementary school children to handle."

This is the most ridiculous state of affairs in the whole absurd realm of schoolroom arithmetic. Actually, textbooks approach division more sanely than they do any of the other three operations. They customarily begin at the big or business end of every division question. Even that heavily handicapped tenth-century monk had the sense to start there, concerning himself first with the number of fifteens in six thousand rather than automatically focusing on the number of fifteens in fifty-two. Unfortunately, he and his successors down to and including contributors to *The Arithmetic Teacher* seem to feel an irresistible compulsion to go all the way from what matters most to what matters least without ever considering the question of the need for this arduous trip.

Division can be the opposite of "the most difficult arithmetic task." For anyone who enjoys guessing games, such as Twenty Questions, it can even be enjoyable, startling though that may seem. Division *is* a guessing game.

Perhaps the strangest part is that it could be treated exactly that way in the classroom without changing much besides attitudes—primarily the attitudes of teachers and compilers of textbooks. The textbooks already present division as a series of guesses, though they call them something else. They call them by the heaviest, dullest, most deadly name they could possibly choose. They call them Estimated Quotients. One would need only a touch of paranoia to suspect a conspiracy to make children suffer instead of letting them have a little fun.

One is not aided in resisting this suspicion by the fact that although the old-fashioned Latinate terms have been dropped from most textbooks in connection with other arithmetic operations, they are still retained in division. Numbers to be multiplied almost never are called multiplicands and multipliers any more, and most of us have been spared from subtrahend and minuend in subtraction and augend and addend in addition. But almost all textbook discussions of the division operation still insist on those ancient clinkers: dividend, divisor, and quotient.

The only contribution of these terms is to confuse matters. Forget them. Forget, if you can, that you ever knew of their existence. They are false companions, intent on leading you into a Dismal Swamp of meaninglessness.

All you need are the simple, sensible, modern terms used in multiplication. The numbers to be multiplied are factors and the result of multiplying one by the other is their product. Division is in order when you know one factor and the product but not the other factor. The way to find that other factor is by guessing at it.

You do not have to guess in the dark, of course. Just by looking at the product and known factor you can tell something about the unknown one. How much guessing you do depends on how much more you want to know.

In textbooks, division is treated as if it were of two radically different kinds—short and long. Short division is so called because the known factor is short, only a single figure or, at most, not greater than the number twelve. This is because the holy multiplication table as taught in most schools stops at twelve times twelve, and you are supposed to need instant command of the table in order to do short division. In the sensible, guessing approach to the operation the distinction between short and long disappears, and the multiplication table steps down off its pedestal.

Textbooks order children to line up the operation of dividing, for instance, forty-seven hundred and nineteen by nine, like this:

$$9\overline{)4719}$$

This is a fine way to set it up if the purpose is to make things easy for a teacher. Often there is an excellent reason, not to say dire need, for a teacher to want to make things easy. He or she may have to grade the test papers of thirty-odd children performing a dozen or so such operations not only for correctness but also for neatness and whatever else the local school authorities happen to consider virtues. All of which helps explain why the teaching of both plain arithmetic and New Math is in such a rut, but it does not make the result any better for the victims.

When your purpose is to find out something you want to know in adult life, setting the operation up that way is clumsy and time-wasting. Suppose you have volunteered to take charge of canvassing your town on behalf of a candidate you strongly favor for a certain political office. You know from the records that there are forty-seven hundred and nineteen registered voters in the town. Eight other persons have volunteered to help you canvass. What you want to know is whether the nine of you can hope to cover all, or nearly all, the voters, or whether you will be so swamped that your first concern must be to try to find more volunteers.

This is one of those many cases in which the words are better than the numerals. The only words to the point here are the nine of the known factor and the forty-seven hundred of the product. (Note, incidentally, that if the product were written 4,719, one might easily confuse the issue by thinking of it as four thousand seven hundred and nineteen.) The "hundred" in the product tells you immediately that there is going to be a "hundred" in the factor you are seeking. The only way you can get nine up into the hundreds range is by multiplying it by one or more hundreds.

The remaining question is by how many hundreds must you multiply nine in order to boost it to the neighborhood of forty-seven hundred. If you are one of those with the multiplication table firmly stuck in your mind, the guess may seem to pop up as soon as you realize what you want to know. Or it may not. Forty-seven is not an exact multiple of nine, and the way the multiplication table usually is learned can make it difficult to recognize near misses.

If the guess does not spring to mind immediately, you may want to use the chart suggested in the chapter on multiplication. Or you may prefer to think of nine as about ten. It is a simple matter to guess at the factor by which to multiply about ten in order to come reasonably close to forty-seven.

But a little consideration of the way we make such a seemingly simple guess helps clarify how guessing goes and how much of it we can do without being fully aware we are doing it. You may feel that you instantly guess that five is the factor which multiplies ten to yield the product closest to forty-seven. In all likelihood it actually is your second guess. You know that forty means four tens, so it is almost inevitable that you think, however fleetingly, of four as the possible other factor. That the factor you want is five, which multiplies ten to yield fifty, which is closer than forty to forty-seven, probably is your second guess. The reason for pointing this out is to make clear that the kind of guessing to be done in division is a natural, relaxed way of doing things. There is no need to be right the first time because you can change your guess as soon as it proves unsuitable.

First, second, third, or whatever guess it may be, five or about five is the number that matters in this case. Link it up with the hundred, and you are well on the way to the answer to your question. You and your eight friends will have to canvass about five hundred voters each if you want to reach almost all of them. A more precise figure would only distract you from the further important matters to be considered in your decision, such as how many days remain for canvassing and how much time you and your fellow volunteers can put in.

To see how close to the mark any guess comes, all you need do is multiply the guessed factor by the known factor. This suggests a good way to write down the numbers involved if you prefer to have them on paper rather than to do the guessing and checking mentally. Instead of lining up the product and the known factor in the textbook way, write them separately. Then write your first guess at the unknown factor near the known factor. When you want to check your guess, write somewhere near the product the round-number result of multiplying the known factor by your guess at the unknown one. Don't care if your scribbling is good and messy in defiance of schoolroom habits:

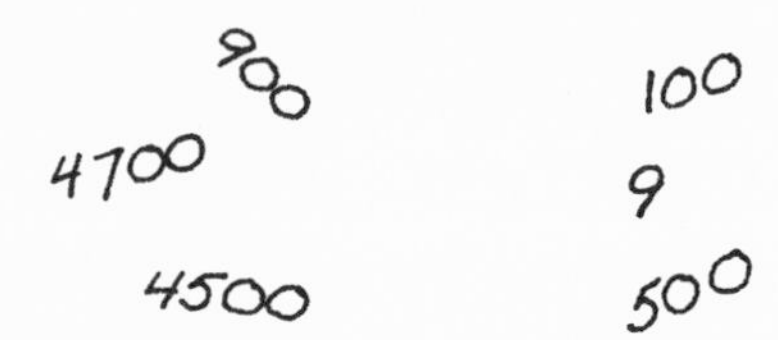

Remember that by themselves the figures are meaningless. They are dependent on you for their meaning. You can do with them anything that suits your convenience or helps you answer the question you have in mind.

In most schoolrooms the kind of division you would do in a situation like that under discussion would stay "short" if you recruited one, two, or three more canvassers—up to a total of twelve—but would abruptly become "long" if you had the luck to find four more and a total of thirteen. In the approach suggested here it is as easy to guess at about how many voters each of thirteen, fourteen, fifteen, or more canvassers will have to reach as in the cases of twelve or fewer.

Indeed, this kind of division can be short and easy when the known factor is in the hundreds. Suppose you decide that there is only enough time for each canvasser to reach about two hundred voters. You want to know how many canvassers you need to have enough to reach the forty-seven hundred voters. Now all you have to guess is by how many you must multiply two to raise it to forty-seven, and you quickly learn that you need a total of twenty-three or twenty-four volunteers.

There are, of course, situations in which you want more details about the unknown factor that multiplies with nine to make forty-seven hundred and nineteen. Perhaps you have helped a group of nine children collect old newspapers and sell them for recycling. The sale produces a total of $47.19, and the children want you to tell them how to divide the money.

In this case the hundreds are hundreds of cents, and one hundred cents make a dollar. You can see—either by consulting the multiplication table in your mind or on paper or by multiplying five by ten and then subtracting one five—that giving each child five dollars will account for $45. This leaves $47.19 minus $45, or $2.19 to be divvied up.

You know that there are four quarters in a dollar and, therefore, eight quarters in two dollars. So you can see that $2.19 does not quite make a quarter apiece for your nine young friends. You might consider putting up enough from your own pocket to stretch the 19 cents into a ninth quarter. Or you might suggest they contribute the $2.19 to some cause. Or that they blow it on an ice-cream-and-cookie spread. That is to say, these possibilities might occur to you if you are not so deep in thrall to the textbook way of dividing $47.19 by nine that you are incapable of thinking of anything while doing it.

To anyone still in that thrall, of course, the way this example has been handled might seem scandalous. Or at least, it probably would seem like a weaseling out at the end, a refusal to face the need to divide $2.19 by nine in a way that accounts for every penny. Perhaps the best thing about the guessing approach to division is that it gives you the chance to notice alternatives to pursuing the process to the bitter end—but guesses can lead you there if you insist. When you notice that $2.19 is six cents short of enough to provide each child a quarter, you also are noticing that it is enough to provide them 24 cents each with 3 cents left over in the pot.

This brings you up against the bane of millions of children—a remainder. Those millions of children are the ones who have been forced to spend many agonized hours on the question of what to do with such remainders. There are two great and deeply divided schools of thought on the matter among teachers of arithmetic. One school holds that it should be written as a fraction (that is, as $3/9$ in the case we have been considering). The other insists on writing it preceded by the letter R for Remainder (in this case, as R3).

This is another of the absurd results of the mechanistic approach to arithmetic. When you focus on what you want to learn with the help of the division operation instead of on the mechanics of performing it, what to do about anything left over is obvious. In the case we have been considering there were several possibilities, such as adding 6 cents to provide each child exactly $5.25, or contributing the $2.19 to some cause; and if it came down to each child having $5.24 with three pennies left over, they could draw straws for the pennies. If you buy one of an item that sells two for 19 cents in a store, you would have to be remarkably out of touch to think that the operation $2\overline{)19} = 9\frac{1}{2}$ or 9R1 tells you what you must pay. In the solution to a problem like dividing seventeen people to be transported somewhere in three cars there is scarcely any resemblance between the obvious natural solution of six in each of two cars and five in the third car and the mechanical $3\overline{)17} = 5\tfrac{2}{3}$ or 5 R2.

There are, to be sure, cases in which there actually is an awkward remainder, but the awkwardness is social or otherwise outside the realm of arithmetic. If there are ten people on hand all anxious to play bridge, two of them simply have to kibitz or go find two more players. This may cause hard feelings, bridge players being what they are, but for such problems arithmetic can provide only advance warnings, not solutions.

One standard textbook way of introducing young victims to the type of division labeled "long" avoids even the washed-out kind of guessing called Estimating Quotients. Given the task of dividing 67,489 by 229, children are told to line the figures up as in short division, then start subtracting 229 from 674, like this:

$$\begin{array}{r} 229\overline{)67489} \\ \underline{229} \\ 445 \\ \underline{229} \\ 216 \end{array}$$

This is supposed to show them that 229 will "go into" 674 twice. They write 2 above the 4 of 674, like this:

$$\begin{array}{r} 2 \\ 229\overline{)67489} \\ \underline{229} \\ 445 \\ \underline{229} \\ 216 \end{array}$$

Then they "bring down" the 8 from 67,489, tack it onto the right end of 216, and start subtracting 229 from the resultant 2168:

$$
\begin{array}{r@{}l}
 & 2 \\
229\overline{)} & \overline{67489} \\
 & \underline{229} \\
 & 445 \\
 & \underline{229} \\
 & 2168 \\
 & \underline{229} \\
 & 1939 \\
 & \underline{229}
\end{array}
$$

And so on.

This is quite literally machinelike, being exactly the way desk calculators work in doing division. In a slightly more humane approach the teacher writes on the blackboard the results of multiplying 229 by numbers from two to nine:

$$
\begin{array}{rrrrrrrr}
229 & 229 & 229 & 229 & 229 & 229 & 229 & 229 \\
\underline{2} & \underline{3} & \underline{4} & \underline{5} & \underline{6} & \underline{7} & \underline{8} & \underline{9} \\
458 & 687 & 916 & 1145 & 1374 & 1603 & 1832 & 2061
\end{array}
$$

The idea behind this is to make the children perform not as imitation desk calculators but as somewhat more sophisticated scanning devices. They are supposed to line up the 229 and 67,489 as before, then set about finding on the blackboard the proper multiples of 229 to subtract from, first, 674, then from 2168, and so on.

In theory the teacher helps the children see the connection between these mechanical performances and the division of 67,489 by 229. In practice millions of children learn to imitate calculators or scanning devices without a glimmer of insight into what they are trying to do. All of which should make it clear why "learning to work division problems is probably the most difficult task for elementary school children to handle."

The humane way of dividing by guessing at the missing factor starts by considering the whole product. It is clear that 67,489 is in the tens of thousands. The first guess to make is at by how much it is necessary to multiply 229 to raise it to the tens of thousands level. Your first, half-conscious guess might be that you must multiply 229 by a thousand. But you quickly realize that a thousand times 229 yields 229,000, which is in the hundreds of thousands and too big. So you step down to one hundred and note that one hundred times 229 yields 22,900, which is on the same tens of thousands level as 67,489.

Next comes the guess at by how many hundreds 229 must be multiplied. You can simplify this guess by thinking of 229 as about 230 and of 22,900 as about 23,000. You might first guess at two and find that two times 230 yields 460, then at three and find that three times 230 yields 690. This tells you that 300 times 230 yields 69,000, a little bit over a thousand more than 67,489. Does missing by that much matter?

It probably does not if you are looking at a building lot with the thought of buying it to build a new home and the agent showing you the lot can tell you only that it is rectangular, covers 67,489 square feet, and has a road frontage of 229 feet. He says he thinks it is about 300 feet in depth. Your guesses show you that he is right.

But suppose that you are very fond of trees, especially

ancient oaks, and you notice a magnificent one well back from the road front. By pacing from the road you find that it is about 280 feet to the oak. (I go into the subject of pacing in Chapter 8.) You feel that you have to know whether it is in the lot, because if it is not and should be cut down by whoever owns the land where it stands, you would be much less interested in buying the lot.

You have found that 300 times 230 yields 69,000 square feet, so the exact depth of the lot is a little less than 300 feet. You want to know whether it could be as much as ten feet less. First, you note that the width of the lot is 229 rather than 230, so 69,000 is one times 300 too high a figure. It is easy to take 300 from 69,000 and find that a lot 229 feet by 300 feet would cover 68,700 feet. This overshoots the 67,489 figure by about 1200. Your question is how many feet you must trim off the 300 to account for that 1200, and you see at once that trimming off ten feet would be far too much. Ten times 229 yields 2290, far more than the 1200 you have to take off. In fact, 2290 is almost twice 1200, so about five feet is all you have to trim off the 300. So the depth of the lot is close to 295 feet.

Don't slip back into textbook perfectionism and demand of yourself that your guesses be perfect. A guess should, by its very nature, have the privilege of being wrong. In the division operation all you ask of a guess is that it help you get a figure the effect of which can compare with the effect you seek. No matter how far off the mark your first one may be, it helps you find out at least a little of what you want to know.

In everyday life the biggest products and known factors most of us usually encounter are on the order of 67,489 and 229. But figures in the millions and higher can be dealt with in the same way if they turn up. If you have a Great-uncle Bim known to be loaded with $9,781,439 and blessed with 117 living relatives, including yourself, and you want to daydream about possibilities, you can see that multiplying 117 by ten thousand yields 1,170,000 and that nine ten thousands will overshoot the 9,781,439 mark by a bit. So your possibilities come, before taxes, to something less than $90,000. If you want a more exact figure, you are trying to escape to a fantasy world and will not be further encouraged here.

An extra added advantage of the guessing-game approach is that you see clearly at each stage the region of numbers in which you are looking for the unknown factor. This is because you start the game by deciding whether the number you are seeking will be in the tens, hundreds, thousands, or higher. A common result of the textbook method is the misplacing of the answer one or more places to the right or left of where it should be, so that it comes out ten times too big or too small, sometimes even a hundred times too big or too small.

Children taught to perform division mechanically have another difficulty not likely to be encountered in adult life, but it is worth knowing about. A child can have firmly entrenched in mind the answer four to the command to divide twelve by three but be utterly bewildered if asked to assign twelve people to three cars with an equal number in each car. Such a victim of mechanization simply cannot see a connection between making such an assignment and dividing twelve by three. One of the virtues of the New Math program is that it has at least noticed this difficulty, although trying to avoid it by training children to chatter about "sets" seems small progress, if any. Giving children objects they can arrange and rearrange—say, twelve blocks to be arranged in three equal piles—is becoming a fairly common practice, however, and is much more to the point. It is precisely because the division problems adults encounter usually involve concrete questions rather than abstract numbers that this difficulty is not a common one in adult life.

Because so many of the numerical problems they are expected to deal with are abstract, children sooner or later come up against the peculiarities of trying to divide zero or to divide with it. This consumes many hours and causes much anguish in elementary schools, especially those with bad cases of New Mathitis. Some textbooks try to make their victims believe that one (or two or one-half or seven hundred nineteen or any number at all) divided by zero equals infinity because zero "goes into" one or any number an infinite number of times. Other textbooks have even wilder and more complicated explanations of the results of dividing by zero or trying to divide it.

It is highly unlikely that you ever will face such a question in a division operation encountered in everyday life. If you want to take twelve children somewhere and no cars show up at the appointed time and place, you do not mess with trying to divide twelve by zero. You either get on the phone, start walking, or give up the trip. And if three cars but no children show up, you do not dither over dividing zero by three. You start looking for the children. And if neither cars nor children show up, no sane person would see the problem as having to do with trying to divide zero by zero.

6. The Parts Department

FRACTIONS

It takes a great deal of time, effort, willingness to accept risk, and plain luck to successfully perpetrate a large-scale theft. In most cases it also takes the cooperation of several thieves, and that leads to the most dangerous part of all—divvying up the swag. The break that law-enforcement men have learned to be most hopeful about is the disgruntlement of some member of the heist who feels he did not get his fair share. The way the subject of fractions is taught in most schools almost guarantees that one or more participants in any such enterprise will feel that way, or can be helped to feel that way by subtle hints fed to newspaper and television reporters. If three getaway cars are used on a job, each with a different driver, and the drivers share equally one fourth of the stolen money, it would be an unusual school experience that helped all three to understand fractions so well that their confidence in the propriety of their shares would remain unshakable. Such a share of even $100,000 or more can seem suspiciously small, especially after it has been spent on a spree.

That they should be unable to comprehend why their share is so small is perhaps one of the few justifications for the mess most textbooks and schools make of common fractions. Only a very few such numbers are at all relevant in the lives of any but a few persons, and the relevant ways of handling them—adding, subtracting, and so on—can be learned rather easily even by many children with learning disorders. The trouble is that most textbooks pay little attention to the truly common fractions. Instead, they go in for endless examples of unlikely ways of operating with exceedingly uncommon ones.

Some things are not fractionable at all. One child is one child, even when she or he has lost an arm or a leg, and never five-sixths of a child or seven-eighths of a child. But many of the things encountered in daily life must be taken apart to be used or understood—pies, cakes, pounds of butter, quarts of milk, yards of cloth, notes of music, hours of time, miles of a trip. And the names of the parts we regularly use of such things are the names of the only common fractions most of us ever have much, if anything, to do with—halves, thirds, quarters, fifths, sixths, eighths, ninths, tenths, twelfths, sixteenths, thirty-secondths, and sixty-fourths.

If it amuses you to add three sevenths to five elevenths or to divide thirteen twenty-ninths by thirty-seven forty-thirds, go ahead. But do not expect to find situations in which you can make any use of such operations—that is, in which elevenths or twenty-ninths or such are fractions of things that matter to you. Do not let yourself be persuaded that you are "sharpening your logic" or "improving your reasoning power," either. You are merely being conned into imitating a calculating machine the way you were taught to do in elementary school. The only sensible reason for performing such operations is that you enjoy doing so.

As the adjective indicates, common fractions are not the only ones. Indeed, they are not even the most common ones, decimal fractions (of which more in a little while) having long since supplanted them for many purposes. Common fractions also have to compete with the sexagesimals, though this is not quite as tough competition as it may sound. The sexagesimals are not sexy but sixtyish. They indicate subdivision of hours of time, degrees of longitude, and such into sixty minutes, and minutes into sixty seconds. We do not usually think of them as fractions, however, and there is no reason why we should.

Common fractions were more usually and appropriately called the vulgar fractions until about a century ago. It takes at least two words or numerals to express each of them—one half or 1/2, two thirds or 2/3, and so on. Whether the bar separating the numerals is horizontal as in $\frac{1}{2}$ or slanting as in 2/3 depends solely on writing or printing convenience.

The two numerals have names, pompous Latin ones as usual. The lower is called the denominator and the upper the numerator. These are so pretentious and since they both end in "ator" they seem so much alike that they have confused millions of people for generations. If you are among their victims, it may help to translate the terms. Denominator means namer and numerator means numberer. In 2/3 the 3 names the fraction as thirds, and the 2 numbers the thirds as two. Or if you would prefer to think of them as uppers and lowers, that's your privilege.

It usually is fairly clear how to go about adding or subtracting fractions of the same name if you think of them in words rather than in numerals. One half plus one half is as obviously two halves as one plus one is two, and two halves of anything are the whole thing if it was not killed or otherwise irreparably damaged by being split apart. Two quarters minus one quarter obviously leaves one quarter, just as two minus one leaves one. But when it comes to fractions of different names, endless textbook rambling about such pointless matters as seven thirteenths plus forty-one fifty-thirds has left much uncertainty about the very simple similar operations that do turn up in real life, such as adding one third to one half.

It is one of the greatest successes in the history of confusion that this should cause any adult the slightest difficulty. Yet millions of highly intelligent people feel that the only way to find the result of this operation is by going into a mechanical dither about a mysterious and elusive something called "the least common denominator." Actually all you really ever need to know in order to perform the operation is that one third is a little less than one half. And all you really ever want to know as a result of the operation is that one half plus one third amounts to a little less than one.

A typical case for this operation in real life would come up if you measured out half a cup of flour, mixed with it salt, sugar, and baking powder, then decided that you wanted to make a bigger batch of whatever you are about. You find, however, that there is only one third of a cup of flour left on hand. It would be a complete waste of time to try to calculate exactly how much the one-half cup and the one-third cup come to so that you can calculate exactly the right amounts of salt, sugar, and baking powder to be added along with the additional flour. Instead, you know that you have now a little less than a whole cup of flour so you add a little less than the amounts of those other ingredients that went into the half cup of flour.

There is a nonmechanical way of getting the exact result of adding one third to one half if you really need such a result. You can do as the Romans did. Common fractions were a branch of arithmetic in which their clumsy numerals actually worked to their advantage in some ways. Lacking the seemingly neat, simple way of writing common fractions that the Hindu-Arabic numerals make possible, the Romans found other ways to visualize and think about most of the more frequently used ones.

Their favorite gimmick was a system of twelfths, from which we got our division of a foot into twelve inches (it also provided our word "inch," the Roman twelfth being called *uncia*). Thinking of them in terms of inches makes it very easy to add or subtract halves, thirds, fourths, sixths, and twelfths – one twelfth being one inch, one sixth two inches, one fourth three inches, one third four inches, and one half six inches. One half plus one third translates into six inches plus four inches, which makes ten inches or two inches less than a foot. You know that two inches are one sixth and that there are six sixths in a whole foot, so one sixth less than a foot is five-sixths, the exact sum of one half plus one third.

One sixth plus one fourth would be two inches plus three inches or five inches or twelfths. One third plus one fourth would be four inches plus three inches or seven twelfths. One third minus one fourth would be four inches minus three inches or one twelfth. And so on.

Fifths and tenths do not fit into this system. But if you have much to do with them in daily life, they probably will be in bottled form, and your interest will be in the contents of the bottles rather than in any arithmetic. Should you insist on being sobersided about the matter, these fractions are best dealt with in decimal form, which will be considered later.

Eighths, sixteenths, thirty-secondths, and sixty-fourths also can be dealt with in decimal form if there is reason for precision about the results of the dealings. But the most common use for these fractions associates them in easily understood ways with halves and quarters and provides a system to parallel the Roman one for those who would like to use something like it with these other numbers. That most common use is in music, with its whole notes, half notes, quarter notes, and so on. Each note on the list down to sixty-fourths lasts just half as long as its predecessor. This relationship, along with the names of the notes, makes it easy to work out that, for instance, an eighth is one fourth of a half or that it takes eight thirty-secondths to equal a quarter or thirty-two sixty-fourths to constitute a half.

The four shortest notes on the list have pertinent and revealing names in British usage. An eighth note is called a quaver, a sixteenth a semiquaver, a thirty-second a demisemiquaver, and a sixty-fourth a hemidemisemiquaver. These terms emphasize that the differences are so slight that it takes a lot of practice and probably some native ability to distinguish among them. It takes incessant practice and rather rare genetically determined talent to be able to play any instrument with consistent differences in the durations of the fractions of quavers. And in playing some of the horns and other instruments it is highly unlikely that anyone ever has been or will be able to play with real distinctions between, say, demisemiquavers and hemidemisemiquavers at most tempos. The sensible first step in dealing with any very small fractions is to make sure that the differences among them are really noticeable, or at least detectable, in the circumstances with which you are concerned.

Another way of dealing with everyday fractions is by consciously imitating what we do unconsciously in operating with small whole numbers. As has been pointed out, most of us know that two plus two equals four because we learned and accepted this convenient way of rearranging two twos in childhood before being exposed to the inhibiting rigidities of textbooks. If you find yourself having to deal with situations in which it is useful to know the more or less precise results of adding or subtracting fourths and fifths or eighths and thirds or other such uncommon and non-obvious combinations, either make yourself a written list or memorize the results of such combinations.

Finally, if in making such a list or in some other situation you are stuck with a need for the exact results of adding or subtracting common fractions that cannot be handled in the Roman or musical-notation way, you may have to fall back on conjuring up a least common denominator or two. The way to tame this operation is to think of it as a translation. What you want to do is translate fractions with different names into equivalents of the same name.

Consider three eighths plus two thirds. The best way to avoid a big mistake in a machinelike operation such as the least-common-denominator way of adding fractions is by making a commonsense estimate to start with. So start by noting that three eighths is a little less than one half and that one half plus two thirds comes to a little more than one. This means that a little less than one half plus two thirds must be very close to one.

With your flank thus protected, you can begin the translation by multiplying together the lowers of three eighths and two thirds. Three times eight (or eight times three) is twenty-four, which is the name of the fraction into which they are to be translated. (It is on rare occasions when you need to do translations like this that you may have to deal with common fractions of sorts not on the brief list given earlier, but they usually are easy to deal with on such occasions.) Since you have multiplied the three of two thirds by eight, you must multiply the two by the same amount, and this gives sixteen twenty-fourths as the translation of two thirds. Since you have multiplied the eight of three eighths by three, you must multiply the three by the same amount, and this gives nine twenty-fourths as the translation of three eighths. So adding two thirds and three eighths translates into adding nine twenty-fourths and sixteen twenty-fourths, and the exact figure is twenty-five twenty-fourths, or one twenty-fourth more than one. Just what we thought from the outset—very close to one.

If you have forgotten the old textbook terminology as thoroughly as it deserves to be forgotten, you may be scandalized to hear that this is an improper fraction. Go right ahead and forget it again. But fractions less than one do behave differently from fractions more than one in certain operations, though the difference has nothing to do with propriety.

The difference is prominent in the multiplication operation. Ordinarily, we think of multiplying as meaning increasing, usually fast and impressively. His fortune multiplied, her troubles multiplied, their rabbits multiplied. But in the kind of multiplying done in arithmetic the operation results in an increase only if both factors amount to more than one. To multiply any number by one is to make no change. To multiply by zero is to wipe out. To multiply by a negative number has results to be discussed in the next chapter.

What we are concerned with here is the result of multiplying by a fraction greater than zero and less than one. It would save a lot of confusion if this operation were known by a name other than multiplication – say, partake. When you multiply, say, two by one half, you do not increase the two. You take away part of it or partake of it. You take away, in fact, just half of it.

In everyday life we usually are well aware, though perhaps not articulately, that multiplying by a fraction is different from multiplying by a whole number. If you are going to make half a recipe calling for two cups of some ingredient, you don't think of multiplying one half times two. You think of taking half the two cups.

We also usually are clear about what we are doing when both factors are fractions. If one of the ingredients in the above recipe is to be used in the amount of half a cup, you simply take half of that half or one quarter of a cup and are not tempted to complicate matters by trying to think in terms of one half times one half.

By now it may have occurred to you that multiplying by a fraction is like dividing by the number of the lower part of the fraction. And indeed, one half times two—or one half of two—can be thought of as two divided by two. But the procedure is as simple as this only when the upper number of the fraction is one. When the upper number of the fraction is any number other than one, the operation gets more complicated.

The operation $3/4$ times 2, for instance, would be written in the mechanical way as

$$3/4 \times 2/1 = \frac{3 \times 2}{4 \times 1}$$

That is, the whole number is written in the form of a fraction, and you multiply the upper number times the upper and the lower times the lower. If both numbers are less than one, say $2/3 \times 3/4$, you would make it

$$\frac{2 \times 3}{3 \times 4}$$

The fact that it can be performed mechanically in this way is why the operation is called multiplication. In algebra and other branches of higher mathematics it is indispensable to be able to perform such operations mechanically, because you often are dealing with extremely complicated numbers. But to deal with the common fractions of everyday life this way makes them unnecessarily complicated and usually confuses the issue.

Consider that last example, $2/3 \times 3/4$. What you want is two thirds of three quarters. It needs only a moment of thought uninhibited by textbook arithmetic to realize that two thirds of three anythings are two of those things, so that two thirds of three quarters are two quarters.

The mechanical way is unnecessary even when the result is not as neatly obvious as in this case. A need to find, say, three eighths of three fifths is a more complicated need than you ever are likely to encounter in daily life. But if it were to come up, you can see that three eighths is a little less than one half and three fifths a little more than one half, so that three eighths of three fifths means about one half of one half or about one fourth. Going about it the mechanical way, you would find that

$$\frac{3}{8} \times \frac{3}{5} = \frac{3 \times 3}{8 \times 5} = \frac{9}{40}$$

In operating with numbers that combine whole numbers and fractions and thus may seem complicated, the mechanical way only complicates matters further in everyday situations. If you want to know the result of two and one half times two and one half, the machine way is to translate the twos into $\frac{4}{2}$ and to add to each the extra $\frac{1}{2}$ making each $\frac{5}{2}$. Now you go through

$$\frac{5}{2} \times \frac{5}{2} = \frac{5 \times 5}{2 \times 2} = \frac{25}{4}$$

But if you go about it in a more human way, you see that two and one half times two and one half means two times two, or four, plus two times one half, or one, plus one half of two, another one, plus one half of one half, or a quarter—that is, four plus one plus one plus one quarter, a total of six and one quarter.

And then there is the matter of dividing common fractions by common fractions—one of the most difficult subjects in arithmetic if you take the textbooks seriously. In everyday life there is literally nothing to it. We simply never have occasion for it. In theory it could be useful if you happen to forget, say, how many eighth notes there are in a half note. You could find out by setting up a complicated operation like

$$\frac{\frac{1}{2}}{\frac{1}{8}}$$

but it is highly unlikely that anyone ever has done so. The natural way is to look it up in a music dictionary or count on your fingers—one eighth plus one eighth makes one quarter, plus one eighth makes three eighths, plus one eighth makes a half.

Nor does anyone ever need in ordinary circumstances to divide a whole number by a fraction. If you want to know how many quarters there are in two dollars, you do not think in terms of dividing two by one quarter. You know that there are four quarters in one dollar and twice that many in two dollars. If you want to know how many inches in four feet, you do not divide four by one twelfth but multiply twelve by four.

It does not very often happen that anyone needs to divide a fraction by a whole number, either. When the need does arise, what has to be done is clear. In the case of the drivers of the three getaway cars mentioned earlier, it is probable that even such men would be able to understand that each of them was supposed to get one third of the one quarter share allotted to them to be shared equally (that is, to be divided three ways). What would make them so easily confusable about the propriety of this share is that the one twelfth of the total it works out to seems such a small part of that total.

Canceling is another operation with fractions presented in the schoolroom with blind insistence. Whenever the upper and lower numbers of a fraction are evenly divisible by the same smaller number, children must either do such dividing automatically or have their answers labeled wrong. That is, two fourths or $2/4$ is wrong because it also can be stated as one half or $1/2$. In real life two quarter notes are not the same as a half note, and there are many situations (for instance, in a telephone booth) when two quarters are much preferable to a half dollar.

After stunning their victims with a few dozen exercises of the

$$\frac{\frac{37}{73}}{\frac{59}{97}}$$

type,. many textbooks plunge into what they insist is the terribly confusing resemblance between common fractions, ratios, rates, and proportions. What is supposed to be confusing is that the figure $1/2$ can stand not only for a half cup or a half inch or such, but also for the presence of, say, one woman for every two men in a group of people—which is a ratio—and can be a very awkward way of expressing the fact that a certain plane can go one mile every two seconds—which is a rate.

Also, the expression $1/2 = 5/10$ can mean that on a map drawn to a scale of one inch for every two miles five inches represents ten miles—which is a proportion.

But these resemblances are confusing only in the abstract. If in everyday life you ever have occasion to write down a ratio, rate, or proportion in a way resembling the way you would write a common fraction, you would know exactly what you mean by what you write. But such occasions are rare. When you want to consider the ratio of women to men in a group of people or to think about the speed of a plane, you do not think in fractions. And if you want to work out how many miles two and one-half inches of that map covers, you would not write down

$$\frac{1}{2} = \frac{2.5}{x}$$

You would simply double the 2.5.

And that 2.5 brings us to the nowadays far more common kinds of fractions called decimals. Actually, we already have had many dealings with them in this book. All the dollars-and-cents questions considered so far have involved them. That is the clue to the sensible way of approaching them much of the time—namely, in terms of dollars, dimes, and cents.

You can write or think of a dime as 10 cents and a cent as 1 cent, but the forms best for our purpose are .10 and .01. .10 means the same as one tenth or $1/10$, and .01 the same as one hundredth or $1/100$, but the .10 and .01 forms usually are easier to operate with. For instance, .10 plus .10 is the same as 10 plus 10 except that you have to remember the decimal point and make the sum not 20 but .20. And making sure the decimal point is where it belongs is the only thing that makes operating with decimal fractions more difficult than operating with whole numbers.

Inevitably, textbooks have rules to be followed mechanically in order to keep the decimal point in the right place. The far better way is to be clear in your own mind about what you are doing. If you realize when you are adding .10 and .10 that you are adding one dime to one dime, you have no need for rules to tell you that the result must be .20 rather than .02 or 2.0. And when you add .10 to .90 and think of these as standing for one dime and nine dimes, you are in little danger of winding up with .01, .10, or 10.0, which are common classroom results.

In addition and subtraction, this approach makes it possible to treat figures combining whole numbers and decimal fractions exactly as you would whole numbers if you wish to do so. The first step, of course, is to make sure that you really care about the fractional parts of the numbers. It may be better to ignore them or to round them off to the nearest whole numbers. But if you are really interested in the decimal parts of a sum like 13.72 plus 8.39 or in what's left when you take 8.39 from 13.72, you have a choice of several ways of thinking about these numbers. You can treat 13.72 as thirteen hundred and seventy-two cents or as one hundred and thirty-seven dimes and two cents or as thirteen dollars and seven dimes and two cents or as thirteen dollars and seventy-two cents. Your convenience is all that matters.

This also indicates what to do on the extremely rare occasions when you want to add or subtract numbers including decimal fractions in the thousandths (also in the almost-certain-not-to-occur-in-daily-life situations when you have to deal with tens of thousandths or smaller decimal fractions). You can translate 13.726 into thirteen thousand seven hundred and twenty-six thousandths or into thirteen hundred and seventy-two cents and six thousandths, etc.

Textbooks tend to make the treatment of the decimal point in multiplication and division of decimal fractions seem like feats possible only for talented and practiced jugglers of abstract concepts. In multiplication, one way to escape this baleful influence is by thinking of one of the factors in terms of dimes and cents and the other as a common fraction. For instance, you can think of .1 times .1 as one tenth of a dime. .3 times .5 can become three tenths of fifty cents, and since one tenth of fifty cents is five cents, three tenths would be fifteen cents.

This is about as complicated a multiplication-with-fractions undertaking as most of us ever encounter in everyday life. But suppose you should find yourself with a need to know the result of a multiplication like 4.86 times 17.39. Begin by noticing that 4.86 is almost five so that the product you seek will be close to five times seventeen. That is five times ten, or fifty, plus five times seven, or thirty-five—a total of eighty-five.

If you do not feel that that is close enough, note that the .39 of 17.39 is about four dimes, and that five times four dimes comes to two dollars or 2.00. That raises the eighty-five to eighty-seven. But 4.86 is a little more than one dime short of five, so you must subtract a little more than seventeen dimes or a little more than 1.70. This cuts the eighty-seven back close to eighty-five and should be the end of the matter since further adjustments can alter the result only a very little.

It is not much, if any, more likely that you ever will have to perform an operation of dividing in which the known factor is a decimal fraction. But if the need ever arises, you can make the operation simple by treating the fractions as dimes and cents and proceeding just as you would in dividing with whole numbers. In textbooks an operation like

$$\frac{.1}{.1}$$

can be made to seem alarmingly difficult, but if you think of it as "how many dimes in one dime," the difficulty becomes laughable. And

$$\frac{21.93}{6.97}$$

translates into the guessing game "by how much must you multiply 679 cents to get 2193 cents?"

It occasionally happens that it is convenient to translate other common fractions besides tenths into decimal fractions. There is little reason for any difficulty with the common fractions really common in everyday life. Keeping to the coin analogy, it is easy to see quarters as .25, halves as .50, and fifths as two dimes or .20. Eighths are among the rare cases when you get as far out as thousandths if you want precision, the exact decimal equivalent of one eighth being .125.

At first glance thirds may seem a little more difficult, but it is a difficulty only for those conditioned to feel uneasy with "about." The only way you can put thirds into decimals is with an "about." One third is about .33. You also can write it .333 or .333333333333 (and so on forever). And three such thirds come to only .9999999 (etc.), but .9999999 rounds off to a neat 1.00. Two thirds can be written .6666666 (etc.) but is usually given as .67

(and .33 plus .67 make another neat 1.00). Half a third or one sixth is written decimally as .167.

Finally, there is the matter of percentages, which sometimes are written as fractions. We will be going into these later from a different point of view. Here it is necessary only to make clear what percent (or percentage) means because the term can be a little confusing. It means not per penny or per hundredth but per hundred. Six percent interest (which usually, but not always, actually means six percent per year) means six cents for every hundred cents or six dollars for every hundred dollars. Six percent of a population means six of every hundred people. Written as a decimal fraction six percent would be .06.

But .06 percent means something quite different—namely, six hundredths per hundred or six per ten thousand. If you ever find a bank offering loans at such an interest rate, be on guard. There is something wrong somewhere. The interest rate may, for instance, turn out to be .06 percent *per day*, which is no bargain.

7. Down Under

NEGATIVE NUMBERS

One morning a few years ago something extraordinary happened in the course of an arithmetic lesson for third-grade pupils in a school in Weston, Connecticut. The teacher had written on the blackboard:

$$\begin{array}{r} 45 \\ \underline{-17} \end{array}$$

"Now," she said, "we know we cannot take seven from five, so . . ."

A boy named Kye put his hand up and waved it frantically.

The teacher paused. "What is it, Kye?"

"There's a way we *could* take seven from five," said Kye. "Seven from five could be negative two."

The teacher was not flustered or resentful at being interrupted. She was interested. She asked the boy to explain why he wanted to do it his way.

"Well," said Kye, "seven from five is negative two, and ten from forty is thirty. So seventeen from forty-five is thirty and negative two. And that's twenty-eight."

At the teacher's invitation he went to the blackboard and wrote out his way of subtracting:

$$\begin{array}{r} 45 \\ -17 \\ \hline -\ 2 \\ 30 \\ \hline 28 \end{array}$$

This is a new way of subtracting without the clumsy fiction of "borrowing." It is in the schoolroom-mechanical tradition, to be sure, starting as it does with the least important and quite possibly irrelevant parts of the numbers involved. But it is the invention of an eight-year-old. A couple of years later a professional mathematician presented the same procedure in a paper read at a conference on math teaching, but Kye had clear-cut "historical priority," the bread-and-butter and source of such accolades as Nobel prizes for professional scientists and mathematicians.

Kye's last name was not given in an account of his invention published by three admiring experts on the teaching of mathematics, B. S. Cochran, A. Barson, and R. B. Davis, in the March 1970 issue of *The Arithmetic Teacher,* perhaps because his parents wisely feared the effects of publicity. All he got out of it seems to have been the satisfaction that is supposed to be the chief reward of all intellectual innovators, plus the admiration of his teacher and classmates. The latter spent weeks working out details of his invention and a system of notation to go with it. They decided that just as the notation 28 means two tens plus eight ones so the notation $3\overline{2}$ could mean three tens minus two ones. The reason for preferring either form would be convenience.

That an eight-year-old has made a discovery in arithmetic is not what makes this story so unusual, however. Uninhibited young minds are naturally innovative. They have to probe around in order to get a grip on concepts presented to them by older children and adults, and such probing sooner or later leads to discovery of new ways of approaching old subjects.

But the innovative tendencies of children make many adults feel, at best, uneasy and, at worst, determined to put a stop to them. This latter adult inclination used to be especially strong in arithmetic classrooms, where innovation was considered impossible and any evidence of it proof that the child was fooling around. So it still goes in many classrooms. What was truly extraordinary in Kye's case was his teacher's immediate acceptance and encouragement of his idea. One of the best results of the current ferment in the teaching of arithmetic is the growth of awareness among teachers that the old ways are not the only ways.

To those who know something of the history of mathematics, it may seem especially remarkable that Kye's discovery had to do with negative numbers. These long baffled everyone. Some of the most brilliant mathematicians could not accept them. Most of the ancient Greeks simply ignored them. Descartes, the seventeenth-century French genius who laid many of the foundation stones of modern mathematics, was certain that a number representing less than nothing was meaningless. As recently as the early nineteenth century, such numbers were labeled absurd by leading mathematicians and logicians.

A contemporary child would be unaffected by all this, however. Today, negative numbers are common experiences. A temperature of ten degrees below zero is written −10°. The result of a day's trading in a certain stock may be written:

	Open	*Close*	*Change*
XYZ Co.	45	43	−2

If a substitute running back of a football team makes only one appearance in a game and is thrown for a seven-yard loss, his performance is listed as:

Yards gained running −7.

Statements about profit and loss, below-sea-level altitudes, changes in electric-current voltages, and many other phenomena also may involve negative numbers.

There are, of course, rules for operating with such numbers in a mechanical way, but this is one case in which most people are able to ignore such rules in most situations. If that substitute running back were to get back in the game and this time be thrown for a loss of three yards, no one interested in the total of yardage lost by him would be likely to go into anything as complicated as the mechanical process $(-7) + (-3)$. It would be obvious that he had lost seven yards plus three yards, and anyone wanting to write down the total in symbols would make it −10 yards.

If he were to get a third chance and at last make a gain, say of six yards, there would be no need for the mechanical $(-10) + (+6)$. It would be clear that he has wiped out six of the ten yards he had lost. His total contribution would be reduced to a loss of four yards or -4 yards.

If he were to get a fourth chance and make another gain of two yards, his record would go up to -2 yards. And if that play were called back because one of his teammates was offside, there would be no problem about how to subtract the gain. The record would go right back to -4 yards.

And finally, if he then had the horrendous bad luck to run the wrong way on a fifth attempt and wind up with a loss of twenty yards more, you might in charitable sympathy avert your eyes from the field of his ignominy and, if you were keeping track, jot down, with a sigh, -24 yards. It is conceivable that, on looking up, you would find the referee returning the ball to its place at the start of that last humiliation. Some infraction of the rules by the other side has wiped out the play. So now you would want to subtract that last -20 from your -24 total. There would be no need for dithering over anything like $(-24) - (-20) = -24 + 20$. You would know that the twenty-yard loss had been, like the whole play, wiped out and that the runner's record was back to -4.

This is the sensible way to go about adding or subtracting negative numbers to or from positive numbers or other negative numbers. Just think of the positive numbers as gains (of yards, dollars, degrees of temperature, or anything else you wish) and of negative numbers as losses of the same.

Then:

1. If there are only one gain and one loss in the case at hand, subtract the Little One from the Big One. If the

Big One is positive, What's Left is positive. If the Big One is negative, What's Left is negative.

2. If there are more than one of either or both gains and losses, find the sum of the gains and the sum of the losses separately, and as precisely as necessary for your purposes. Then do with these sums as above.

3. If a gain you have included in a total is nullified, treat it as a loss and subtract it from the total.

4. If a loss you have included in a total is nullified, treat it as a gain and add it to the total.

In almost all cases in which you are likely to encounter negative numbers in daily life, all this is likely to be obvious. The only thing to confuse anyone in adding or subtracting such everyday negative numbers is a belief that they are somehow complicated and must be handled according to esoteric rules rather than in a commonsense way.

Unfortunately, it is not so easy to be clear about just what is going on when negative numbers are involved in a multiplication operation. This is because we do not do much, if any, multiplying with such numbers in everyday life. Why bother about it, then? Because, paradoxically, this is one of the very few cases in which operations with abstract numbers can be helpful in everyday situations if such operations can be done with understanding and ease, and not in an anxious and mechanical way.

There are situations in which it is possible at least to imagine multiplying actual gains and losses rather than mere positives and negatives. Suppose (1) you buy ten shares of a stock that rises in value by two dollars per share. It would be easy and pleasant to work out that you have gained twenty dollars, on paper anyway. And if the stock should, instead, (2) fall by two dollars per share right after your purchase of it, you can easily, if not so pleasantly, calculate your loss as ten times minus two dollars or minus twenty dollars. And if you were (3) to sell ten shares of a stock you already own just before that stock rises by two dollars per share, you could think of the result as minus ten times a gain of two dollars or, again, minus twenty dollars.

In other words:

(1) If both factors are gains (or positive), the product is, obviously, a gain (or positive).

If one factor is a gain (or positive) and the other a loss (or negative), the product is a loss (or negative). In (2), above, the ten shares are positive and the fall of two dollars per share in value is negative. In (3), above, the ten shares sold are negative and the rise of two dollars per share positive.

For many people the difficulty is to grasp what happens when both factors are negative. Here is a way to get a grip on this:

Suppose that you were to sell your ten shares just before a two-dollar-per share drop in the value of the stock. You are minus ten shares. You also are minus a loss of two dollars per share. You could think of this as a gain of twenty dollars.

In other words, *a loss of a loss is a gain*, or a minus times a minus is a plus.

To see how multiplying with negative numbers can be helpful in everyday situations, hark back to the chapter on multiplication and the suggestion about how to find out whether forty-four buses with a capacity of sixty-three passengers each would be enough transportation for a certain number of people. The basic suggestion was to start with the round numbers forty and sixty. This product will tell you all you need to know if the number to be transported is only about 2000, the product of forty times sixty being 2400. But if the number to be transported is about 2800, you need to know more. To learn more you take the forty-four times sixty-three apart and make it forty plus four times sixty plus three, then work out as much as you need to know about the four products:

40 times 60
plus 4 times 60
plus 3 times 40
plus 3 times 4

Now suppose that you still have the forty-four buses but that the capacity of each bus is sixty-nine passengers and the number of people to be transported about 2800. You can handle these numbers as you did the others and work out what you need to know about the sum of the four products:

40 times 60
plus 4 times 60
plus 9 times 40
plus 9 times 4

But this has the drawback that the first product is far off the mark. Forty times seventy would be much closer, because sixty-nine is much closer to seventy than to sixty. If you are confident about the results of multiplying with negative numbers, you can round off upward from sixty-nine to seventy, because when you need to know more than the product of the round numbers forty times seventy, you can think of the sixty-nine as seventy minus one. The four products resulting from multiplying 40 plus 4 times 70 minus 1 are:

40 times 70
plus 4 times 70
minus 1 times 40
minus 1 times 4

If the number of people to be transported is about 2800, it is clear that you have more than enough buses. If the number is about 3100, it takes only a little longer to note that you may need another bus or a little crowding.

But this procedure is not indispensable. If multiplying with negative numbers makes you uncomfortable, don't bother with them. The only reason for using them is convenience.

As for dividing negative numbers or dividing with them, occasions for wanting to do either are extremely rare. If such an occasion should turn up, remember that you are looking for a missing factor and that it can be negative, or a loss, only if either—but not both—the known factor or the product represents a loss. If the product represents a loss and the known factor a gain, then the unknown factor must be a loss. If the product is a loss and the known factor a loss, the unknown factor must be a gain. And if the product is a gain and the known factor a loss, the unknown factor also must be a loss.

Finally, if you find yourself dealing frequently with negative numbers and like to write them down, here is a simple and valuable suggestion from the late Max Beberman, director of the University of Illinois Committee on School Mathematics Projects and one of the most effective leaders in the drive to reform the way arithmetic is taught in schools. One of the most confusing things about negative numbers is that the symbol for them is the same as the symbol for the subtraction operation. Beberman suggested making it different by writing it smaller and at the numeral's upper left—for instance, instead of -1 make it ${}^{-}1$. Although the inertia of the old ways of doing such things is hard to overcome, this is being seen more and more frequently, and it probably will be adopted everywhere in the long run.

8. Back to Natural Measuring

ESTIMATING

"Measurement is one of the notions which modern science has taken over from common sense. Measurement does not appear as part of common sense until a comparatively high stage of civilization is reached; and even the common-sense conception has changed and developed enormously in historic times. When I say measurement belongs to common sense, I only mean that it is something with which every civilized person today is entirely familiar. It may be defined, in general, as the assignment of numbers to represent properties."

So wrote Norman R. Campbell, a British physicist and popularizer of science, in his excellent book *What Is Science?*, first published in 1921 and still one of the best on the methods of the physical sciences. His words about measurement evoke automatic agreement in almost everyone who has survived a few years of elementary and secondary schooling, because they are echoed endlessly in textbooks. They still are an unchallenged part of the arithmetic catechism. In the May 1971 issue of *The Arithmetic Teacher,* for instance, an article about the results of the latest research on how to introduce elementary-school children to the concept of measurement informed teachers that "Measurement is a process whereby a *number* is assigned to some object."

This is extremely misleading. Campbell himself apparently felt a little uneasy about his paragraph quoted above. He went on to remark that "only some properties and not all can be thus represented by numbers." But for him and his fellow devotees of science this seems to mean that properties that cannot be represented by numbers cannot really be measured at all.

The kind of measuring Campbell refers to is only one of three different kinds in which human beings regularly engage. In daily life the scientific kind is the least frequently used and of least importance except in the minds of those obsessed with the notion that scientific truths are the only truths.

Measuring, to quote Webster's New International Dictionary, Second Edition, means ascertaining "the extent, degree, quantity, dimensions or capacity of, by a standard." Scientific measuring does indeed involve answering such questions as how long, how much, how fast, how bright, how hard, how dense, etc., with detailed numerical statements that can be checked by others. A typical such measurement would be stating the length of a certain bar of metal as 1.037 meters, assuming that your measuring devices and procedure makes it possible for you to be sure that the actual length lies somewhere between 1.0365 and 1.0374 meters. If others using the same devices and procedure report the same result, the measurement would everywhere be accepted as truly scientific.

A second kind of measuring involves answering some of the same how long, how much, etc., questions sometimes with more or less approximate numbers but often with no numbers at all. Is there enough toothpaste on my brush? Did I put the right amount of sugar in my coffee? Have those wool socks shrunk too much to fit me? We can even do this kind of measuring without being fully conscious of doing it, as when we reach for something, take a step, or notice that some person or object in a dream is strangely large, small, lopsided, or whatever.

A third kind of measuring is by far the most important to us and has nothing at all to do with numbers. It is our own measurement of our subjective states, such as happiness. Students of behavioral science usually prefer to treat such states as not measurable at all, but most of us have memories of many stages of feeling ranging from abject misery to ecstasy or thereabouts. Such memories provide us with standards for making quite fine distinctions.

"Did you enjoy yourself this evening?"

"Well, it was a pleasant enough party. Not so interesting as that fascinating dinner at the Smiths' last week. But far better than that dull cocktail party at the Joneses'."

Obviously, each of us must do this kind of measuring for herself or himself. Just as obviously, it has nothing to do with arithmetic, humane or otherwise. What we are concerned with here are those occasions when we are engaging in the second kind of measuring and have use for more or less approximate numbers.

The reason for going into the other kinds of measurement is to try to pry the whole subject loose from the textbooks and reclaim it for common sense. Presumably not even the compilers of textbooks would argue that it is necessary or desirable to measure out toothpaste on a brush in a certain amount accurate to within a certain fraction, decimal or otherwise. But their teachings about measurement have helped to make commonplace other absurdities at least as great as this, such as wrist watches "accurate to one second per day" and selling for hundreds of dollars to thousands of people who never need to know the time more accurately than within a few minutes of exactitude.

Textbooks have helped make possible such absurdities because they have persuaded millions of people that measuring always must be done with care and that the best measurement is the most precise one. Many people emerge from school stuck for life with belief in the desirability and attainability of perfect accuracy in measuring. In reality, the only perfectly accurate measurements are those that involve counting separate and indivisible items such as people, eggs, and railroad cars. No matter how "scientifically" it is made, a measurement of 1.0 foot means somewhere between .95 and 1.04. A measurement of 1.00000 means somewhere between .999995 and 1.000004.

Of course there are measurements that have to be accurate or fairly accurate. If you're a pharmacist, for instance. But in most everyday life situations accuracy closer than "about one foot" would be pointless. It is wasteful of money to buy—and of time to use—rulers or other such instruments so made as to provide unnecessarily precise results. And one of the worst aspects of the textbook doctrine on measurement is that it either insists on or takes for granted the use of some sort of manufactured measuring gadget even when all that is needed is a rough approximation. This has left millions of people helpless when such gadgets are not available and completely ignorant of the measuring instruments they were born with.

Our natural equipment provided the standards for doing much of our measuring before "modern science took over from common sense," and for many purposes that equipment still is far superior to anything you can buy in a store. One of the most useful items already has been sneaked into this book. You may remember that in the chapter on division it was assumed that the depth of a plot of land could be measured "by pacing from the road." This was sneaky because in these over-mechanized days probably not one person in a hundred is able to make a reasonably approximate measurement with his or her pace. When eleven women of various ages and occupations recently were asked about the length of their paces, only two could remember ever having given the matter any thought. Twelve of fourteen men had at least thought about it, but each of the twelve believed his pace to be considerably longer than it actually proved to be.

In a great many situations your pace is the best possible standard for measuring lengths, so it is worth getting to know its approximate value. But first, be clear about what a pace consists of. The reason why many people overestimate the pace length is that they think of it as stretching from the heel of the rear foot to the toe of the front foot. To see why this is wrong, consider how you would measure the next pace. If you start from the heel of what was the front foot and is now the rear one, you would be measuring the length of that foot a second time. The ground actually covered in each pace is that stretching from heel of rear foot to heel of front foot or from toe of rear foot to toe of front foot, whichever you find easiest to start with in a given situation.

One way to proceed is by concentrating on your ordinary pace. Put a yardstick or tape measure on the floor at a spot where you can approach it walking naturally, then stop alongside the stick with both your feet flat on the floor. Or try to walk naturally, stop with both feet flat on the floor, and have someone else measure the heel-to-heel distance. Keep at it a few times until you are sure you have a good idea of about how far you usually stride.

The figure varies a good deal from person to person, depending on height, leg length, and walking habits. For women of about five feet two inches in height, it usually is about twenty inches. For most men of about five feet eight inches in height, it is about twenty-six inches. For most men of about six feet in height, it is about thirty inches.

But there is another kind of stride that can make the arithmetic much simpler. Instead of using your ordinary way of walking, create for yourself a special pace for measuring purposes. Most people can do this by stretching the natural pace a little. A good stretched pace for smaller women is about two feet in length. Taller women can use two and a half feet, and most men can stretch to three feet. A good way to practice a stretched pace is by marking a ten-yard stretch of ground and pacing it off a few times.

The second most useful measuring standard is the distance from tip of thumb to tip of little finger with the hand stretched to its widest. For many women this is about seven inches and for most men about nine inches, the latter being the figure most dictionaries give for the length of a span. This can be extremely useful in self-service stores, where there is likely to be little or no chance of getting someone to help you determine the diameter of a frying pan, the height of a desk lamp, and other such measurements.

The term "hand" still is used in stating the height of a horse at the shoulders, one hand being four inches. This is about the actual width of the average man's hand from the base of the thumb across to a spot below the base of the little finger, the thumb and fingers being kept together. For the average woman the hand's width is about three and a half inches. The width of an average man's thumb at the base of the nail is about one inch and of an average woman's thumb about two-thirds of an inch.

Another measurement still sometimes used is the one from nose tip to fingertip with the arm stretched out to the fullest at the side. For the average man this is about one yard, for the average woman a few inches less than that. It is very convenient for measuring cloth and other things that can be held against the nose with one hand and stretched out straight with the other hand. Along with this goes the cubit, the distance from elbow to fingertip, which is about half the nose tip to fingertip distance.

Incidentally, and just a little bit contradictorily, learning the natural measurements offers an opportunity for Britons and Americans to get used to the metric system, which now is in use throughout the world. The British government began making its use official in 1965, and many U.S. industrial firms with foreign customers are using it unofficially. At the beginning of 1972 the only governments still holding back from making it official were Barbados, Burma, Gambia, Ghana, Jamaica, Liberia, Muscat and Oman, Nauru, Sierra Leone, Southern Yemen, Tonga, Trinidad, and the U.S.A.

You can make your special measuring pace, for instance, doubly a step in the right direction by thinking of it not as about two feet in length but as about sixty centimeters, or not as three feet but as about ninety centimeters.

For volumes and weights we have no such direct natural measuring standards as the pace and span, but it is easy to learn to make close approximations to small ones. It is easy, that is, once you shake loose from yearning for exactitude. If you are irritated or otherwise put off by recipes that call for a pinch of this or a dash of that instead of one eighth of a teaspoon of this or one tenth of an ounce of that, you still have a bad case of textbookitis. To feel that you must measure with gadgets like teaspoon fractions instead of by what you can pinch up between thumb and forefinger is to feel that it is better to be as much as possible like a machine, which is what textbooks preach. The difference between a pinch and an eighth of a teaspoon is the difference between spontaneity and rigid conformity.

Volumes and weights larger than those used in cooking are more difficult to approximate, but then, there seldom is any reason to want to guess them. Usually, though, it is possible to tell which is heavier of two liftable things if one of them is at least 10 percent heavier than the other. And if there is any need to do so, most people can learn to guess within about 10 percent the weights of things they can lift.

Two natural standards for measuring time are readily available to all. One is the pulse, which averages seventy to seventy-five beats per minute. It varies a little from person to person, gets a little slower as you grow older, and is faster after exertion or strong emotion. You can easily get the tempo of your own. If you have trouble feeling it in your wrist, try the side of your throat.

This pulse was the timing standard young Galileo used in making the famous first of his long series of discoveries about the laws of motion. Sitting in a cathedral, he noticed a lamp swinging back and forth at the end of a long cord. He also noticed that the length of the swing grew steadily less and wondered whether the time of each swing also diminished. Using his pulse for timing, he found that the time of a complete swing remained the same. That is, no matter how far the lamp swung, it took the same number of pulse beats to get back to whatever starting place he chose to begin counting at.

The other natural standard for measuring time is provided by the earth's daily rotation on its axis. The axis on which the earth revolves is pointed toward the Pole Star, and the direction of revolution is toward the east or, in the northern hemisphere, clockwise. Consequently, the star constellations easily visible in the northern hemisphere, such as the Big Dipper, seem to revolve counterclockwise. Just think of such a constellation as the hour hand of a twenty-four-hour clock moving opposite to the usual direction. To be able to tell the time of night from this clock all you need to know is the position of the hand at some given time, say the first time you glimpse it after sunset, and fix it in your memory by relating its direction to something on the ground. Later the angle it has moved away will be the same angle as the hand of a 24-hour clock would have moved, except that it will be in the opposite direction if you're in the northern hemisphere.

In the southern hemisphere things are even easier. When you look toward the point in the heavens toward which the South Pole points, the earth is revolving counterclockwise, and the constellations seem to revolve clockwise.

To tell the approximate time of day from the position of the sun, it helps to have a few bits of information. In the middle half of a day, you can make do with no more than the knowledge of the direction of due south (or, in the southern hemisphere, due north). The sun crosses the line running from the zenith to the due south point on the horizon (or due north point in the southern hemisphere) at noon, so its distance from the line tells you pretty well how near noon is.

To make reasonable approximations early or late in a day you also need to know about where and when the sun rises and sets that day. This makes it possible to visualize the path it follows from sunrise to the noon meridian and from that meridian to sunset. With the path visualized, it is easy to see how much of it has been traveled or remains to be traveled.

All this, however, gives you only what might be called the real time of day. Clock time may be a good deal different. To translate to it you may have to take into account your position in your time zone. If you are near the eastern edge of the zone, clock time will be later, and if near the western edge, earlier.

And then there is Daylight Saving Time. The whole idea is in the name, the intention being to save daylight for those who like to spend long daylit evenings outdoors. During the part of the year when there is daylight to spare, an hour is taken from the morning and used in the evening. This is done simply by setting clocks ahead one hour. Consequently, while Daylight Saving Time is in effect, clock time is one hour later than real time.

9. The Percentages

PROBABILITY

One day in the early 1650s Blaise Pascal, a remarkable combination of scientist and mystic, was traveling southward from Paris by coach. An aristocratic hanger-on of the royal court shared the coach with him. To pass the time the latter asked Pascal's opinion about a gambling problem—namely, how to split the pot in a certain dice game that had had to be discontinued before it could be finished. Pascal's wide interests included mathematics, and the question intrigued him. Unable to solve it at once, he discussed it at length in correspondence with his friend, the magistrate Pierre de Fermat, who also liked to indulge in mathematics in his spare time. That celebrated correspondence launched the development of the branch of mathematics now known as probability theory.

Since then thousands of mathematicians, including many of the greatest, have worked long and hard on probability studies and have made this one of the most useful mathematical specialties. Indeed, it has become so important to scientific research that historians of mathematics have puzzled over why it did not receive attention until so late a date and have poked around among old manuscripts in search of hints of earlier concern with it. The only notable find is a little book on games of chance by Girolamo Cardano, the sixteenth-century Italian gambler and mathematician, but no other scholars seem to have paid any attention to this book until the historians rescued it from oblivion.

The long postponement of attention to probability theory is no recondite mystery of interest merely to scholars. By far the greater part of the reasoning human beings do not only in science but also in daily life is based on evidence insufficient for certainty. Or, as the seventeenth-century philosopher John Locke put it:

"In the greatest part of our concernment God has afforded us only a twilight of Probability, suitable, I presume, to that state of Mediocrity and Probationership He has been pleased to place us in here."

This suggests an explanation for the lack of interest in the subject among earlier mathematicians. It was too familiar, too obvious, to be worth study, just as it seemed to be obvious that the sun and stars revolve around the earth. People took it for granted that they knew all they needed to know, or anyway all they could know, about probabilities. They knew intuitively. It also helps explain why the first breakthroughs that upset erroneous conceptions were made by study of gambling problems. Coin tossing, crap shooting, roulette, poker, and other such probability problems can be expressed numerically and solved clearly and neatly. Probabilities encountered in everyday life tend to be comparatively vague and messy.

In this case textbooks do not deserve great blame for the general failure to make good use of what mathematics has to offer in the way of help in dealing with everyday problems. Not that textbooks are much help.

Many efforts have been made lately to devise approaches to probability for high-school and even elementary-school pupils, but the texts for such courses take the easy way of concerning themselves exclusively with neat, clear gambling problems and ignoring everyday messiness.

The fundamental reason for failure to utilize knowledge of probabilities, however, seems to be that split between the two sides or aspects of the human brain and mind—the split between the analytical-verbal-logical side and the holistic or intuitive side. Up to this point in this book, much has been made of the need to strengthen the intuitive side. In dealing with probability it is the intuition that requires taming so that the two sides can work together as a team, because a great many of our intuitive notions about probability are mistaken.

In the 1950s a British psychologist, John Cohen, and his associates at the University of Manchester made a series of studies that vividly demonstrated the faultiness of intuition in this area. One study involved comparing the probabilities of similar events, such as the chances of drawing a single lucky ticket from boxes of tickets. Cohen first gave his subjects a choice between drawing a single ticket from a box of ten, or ten tickets from a box of one hundred (putting back the ticket after each draw and shaking up the box before the next draw). This is a choice between one 10 percent chance and ten 1 percent chances. The easiest possible arithmetic makes it obvious that the probability of success is the same in each case. Nevertheless, four out of five of Cohen's subjects preferred the single draw from the box of ten. The same proportion of subjects held to this preference even when they were given the alternate choices of making twenty, thirty, forty, or even fifty draws from the box of one hundred. That is, they preferred the 10 percent chance to chances of 20 percent, 30 percent, 40 percent, and 50 percent. Cohen found that at least part of the reason for this great misunderstanding of the probabilities involved was linked with irrational fear of drawing the same ticket over and over.

But the most startling result came when the subjects were given a choice between one ticket from a box of ten and one ticket from each of ten boxes of one hundred tickets apiece. In this case more than half the subjects preferred the ten draws over the single one. This means that they preferred ten chances in a thousand, a 1 percent chance, to one chance in ten, a 10 percent chance.

In everyday life choices are almost never as plain and numerically obvious as in Cohen's tests, but it is rare for intuitive preference to take any account of any mathematical probability that can be detected. For instance, you are writing a letter at a desk in your living room and suddenly notice that a gemstone is missing from its setting in the ring on your finger. You remember noticing it in its place just half an hour earlier. During that half hour you have had your house entirely to yourself. You cannot remember making any motion likely to have jarred loose the stone. You have spent almost all of the time either at your desk or walking about the living room thinking about what you are writing, the only exception being that you stepped into the next room once to look at the clock, a maneuver that took one minute at most.

Where would you look first for the gemstone? Where would you concentrate your search if the first few moments of it yielded no results? Most of us would start with the next room, if not concentrate there, simply because the memory of having gone there would draw one's attention. Yet from the evidence cited there would appear to be about one chance in thirty that the stone fell out in the next room versus twenty-nine in thirty that it fell in the living room.

To be sure, this is a trivial case. But so are most cases in which we have to choose between different probabilities. A large number of trivialities, however, add up to results that are not so trivial. Also, being in the habit of letting your intuition misguide you about trivial possibilities means that you do the same in more important situations.

Suppose that it is extremely important for you to intercept someone who is en route by car past your neighborhood. The car is due to pass within the next minute or two. There is, so far as you know, equal likelihood that it will travel along either of two streets, one of which passes a hundred yards to the north of where you are, the other a hundred yards to the south. Traffic is heavy on both, and you won't be able to spot the car until it is quite close. Your only way of getting to either street is on foot. What is your best chance of intercepting the car?

For most of us the intuitive course of action would be to dash first to one street and wait there a few moments. If the car does not appear, we would then dash to the other street. If the car does not appear in the second street within a few moments, some of us would dash back to the first one. Since the car may be on either street, we feel that it is useful, or even essential, to try both streets.

But the best chance is simply to go to one street and wait there. This provides a 50 percent probability of making the interception, the maximum possible in the situation described. You have no chance of making the interception while dashing from one street to the other, so any time you spend doing this serves only to reduce the probability of success.

There is no easy way to recognize all such intuitive mistakes about everyday probabilities. It would be a good idea, however, to suspect one's first impulse in such situations, especially when that impulse is to do something that seems to promise a big improvement in the chance of success. Not many people are content with a 50 percent chance or even a 67 percent chance, though in many cases much smaller chances are simply the best available. Such discontent leads to strenuous grasping at deceptive straws.

One of the most grasped-at straws involves a widely and devoutly shared misunderstanding about probabilities. Consider the case of the head of a large family who was planning a homecoming party for his children, grandchildren, and a few nieces and nephews. His country home was roomy, but with so many children to be kept busy it seemed better to have the get-together outdoors on a pleasant day. For various reasons the day had to be either the Saturday or Sunday of two weekends in mid-May—May 12, 13, 19, or 20—and the day had to be chosen two weeks in advance.

He was able to learn from the Weather Bureau that the long-range forecast was of normal precipitation for the area during May and that it normally rained on only one day of three during the month. This means that he would have two chances out of three of picking a rainless day if he chose at random. He was sure he could improve on this probability. To do so he spent most of a day at a newspaper office in the nearby city and was able to get reasonably reliable reports about precipitation on those dates in each of the preceding sixty years. He found that it had rained on twenty-one of the May 12s, nineteen of the May 13s, twenty-one of the May 19s, and eighteen of the May 20s.

What interested him most, however, was that in the preceding eight years it had rained on five of the May 13s but on only two May 12s, three May 19s, and two May 20s. This persuaded him that it was far less likely to rain on the following May 13 than on any of the three alternative dates. He reasoned that May 13 had used up more of its chances of being rainy. So he chose May 13 —and was sadly disillusioned when it drizzled most of the day.

The straw he grasped at was the Monte Carlo fallacy, so-called because it is the basis for innumerable hapless schemes to break the bank there. Almost everyone seems to "know" intuitively that if a chance event occurs a few times in succession the probability that it will occur still one more time is reduced. What this means is clearest in coin tossing.

If you toss an evenly balanced coin, there is a 50 percent chance—or a $1/2$ chance—of its coming up heads. If you toss it twice, the chance that it will come up heads both time is $1/2 \times 1/2$ or $1/4$. If you toss it three times, the chance that it will come up heads all three times is $1/2 \times 1/2 \times 1/2$ or $1/8$. And so on.

According to the Monte Carlo fallacy this means that, if it comes up heads on the first toss, there is less than a 50 percent chance that it will come up heads on the second toss; and if it comes up heads twice in a row, there is still less chance that it will come up heads on the third toss; and so on.

The statement that if you toss a coin twice there is a $1/4$ chance that it will come up heads both times means that there are four different possible results in two tosses and having it come up heads twice is only one of these four. The four possibilities are:

	First toss	*Second toss*
1.	Heads	Heads
2.	Heads	Tails
3.	Tails	Tails
4.	Tails	Heads

Two of the four possibilities, or 50 percent of them, have the coin coming up heads on the first toss. According to the fallacy there should then be less than a 50 percent chance that the next toss would result in heads. But as the table shows, there are two possibilities for the toss following a first toss of heads, and one—or 50 percent—of those two is heads.

A table of the possible results of three tosses may help make this clearer. The statement that if you toss a coin three times there is a 1/8 chance that it will come up heads all three times means that there are eight possible results in three tosses and having it come up heads three times is only one of these eight. The eight possibilities are:

	First toss	*Second toss*	*Third toss*
1.	Heads	Heads	Heads
2.	Heads	Heads	Tails
3.	Heads	Tails	Heads
4.	Heads	Tails	Tails
5.	Tails	Tails	Tails
6.	Tails	Tails	Heads
7.	Tails	Heads	Tails
8.	Tails	Heads	Heads

In four of the eight possible results of the first toss the coin comes up heads, and in two of the four second tosses succeeding these four first tosses the coin comes up heads. According to the fallacy there should be less than a 50 percent chance that a third toss following a first two tosses of heads will result in heads. But as the table shows, there are two possible results of such a third toss, and one of these two is heads.

In everyday life, of course, events seldom are as independent of each other as are the consecutive tosses of a coin. One day's weather has a lot to do with the next day's, and study of such remote phenomena as the stratospheric jet stream can give some indication of long-range possibilities. But in trying to judge two weeks or more in advance the probability of rain on a certain day, you seldom can hope for more useful information than the number of rainy days likely that month in a normal year and how near normal the month seems likely to be in its weather. Information that it has rained on that day in a larger than expected number of the last few years is no more useful than information that a coin has come up heads several times in a row.

"Chance," wrote Joseph Bertrand, a French mathematician, in a famous comment on the Monte Carlo fallacy, "has neither conscience nor memory."

Just as widespread as this fallacy, and sometimes more dangerous, is the habit of ignoring the chance of events of small probability. To be sure, if the probability of an event is extremely small, it may be not only sensible but also essential to ignore that small chance. It would be insane to make your home in a deep mine in order to reduce the very, very small chance of being struck by a meteorite, or to wear knee-high leather boots when walking city streets in order to reduce the danger of being struck by a poisonous snake. But modern civilization is so complex and the number of daily encounters between individuals so enormous that certain somewhat larger, though still quite small, probabilities tend to be become almost inevitable eventualities. This is one of the reasons for the high rate of automobile accidents.

Imagine, for instance, that you are driving along a secondary road and halt at a stop sign. Another car is approaching from the left along the through road. Its directional signal is blinking to indicate a right turn. There are no other cars in sight, and if this one turns into the road where you are stopped, the intersection will be free for you to cross.

If you start out across it before the approaching car begins to turn right, you have a good chance of getting away unscathed. Each time you find yourself in such a situation and react this way you have that same good chance of getting away with it. But if you keep on encountering such situations and reacting this way, there is an excellent chance of an eventual crash just as there is an excellent chance that you eventually will get heads three or four or more times in a row if you keep on tossing a coin. For the probability that the approaching car will continue through the intersection instead of turning right is not at all comparable to the probability of being struck by a meteorite. There is a small but not easily dismissed chance that the right-turn signal is operating without the driver's knowledge. It also is possible that the driver is planning to turn into a driveway beyond the intersection or that he is planning to turn into your road but will change his mind at the last moment. It is impossible to assign numerical probabilities to these alternatives, but it is as foolish as it is easy to get in the habit of ignoring their existence.

Closely similar in intuitive mistreatment of probability is the habit of putting two and two together much too firmly. For instance, Jane tells Jim that Joe is in financial trouble. Jim sees Joe reading a racing form. Jim reasons that Joe's troubles are a result of heavy gambling.

We seldom pause to analyze such reasoning. A twinge of guilt about "spreading gossip" or a righteously generous urge to "give the fellow the benefit of the doubt" may slow the next step of acting on the conclusion, say by passing it along to others as an observed fact about Joe. But this is scarcely analytical. If one really stops to think, it is obvious that putting two and two together about any matter as complex as the behavior of a human being can never yield more than a guess, and that to be aware one is guessing and not deducing certainties in the manner of Sherlock Holmes is to be far wiser, though much less entertaining, than that great sleuth. In the case cited, Jane might have been mistaken about Joe's financial status, and he might have been looking at the racing form out of mere curiosity, to mention only two of the more obvious of the many possibilities that would take the two and two apart.

Unfortunately, there often is great emotional power behind intuitive decisions about probabilities. Strong motivation for interpreting probability in any given way is likely to make that interpretation unshakable. Most smokers who cannot quit are quite sure that they are among the few who can smoke without incurring lung cancer, heart disease, emphysema, or other such consequences; and hypochondriacs need only hear of the symptoms of a rare disease to start finding those symptoms in themselves. About all that can be said is that the capacity to doubt one's own strongly desired assessments of probabilities is worth cherishing.

Index